Springer Tracts in Modern Physics   96

W0079169

Editor: G. Höhler
Associate Editor: E. A. Niekisch

Editorial Board: S. Flügge H. Haken J. Hamilton
H. Lehmann W. Paul

Springer Tracts in Modern Physics

* denotes a volume which contains a Classified Index starting from Volume 36.

S. Büttgenbach

Hyperfine Structure in 4d- and 5d-Shell Atoms

With 14 Figures

Springer-Verlag Berlin Heidelberg GmbH 1982

Dr. Stephanus Büttgenbach

Institut für Angewandte Physik der Universität Bonn, Wegelerstraße 8,
D-5300 Bonn, Fed. Rep. of Germany

Manuscripts for publication should be addressed to:
Gerhard Höhler
Institut für Theoretische Kernphysik der Universität Karlsruhe
Postfach 6380, D-7500 Karlsruhe 1, Fed. Rep. of Germany

*Proofs and all correspondence concerning papers in the process of publication
should be addressed to:*

Ernst A. Niekisch
Haubourdinstrasse 6, D-5170 Jülich 1, Fed. Rep. of Germany

Library of Congress Cataloging in Publication Data · Büttgenbach, Stephanus, 1945 — Hyperfine structure in 4d- and
5d-shell atoms. (Springer tracts in modern physics; 96) Bibliography: p. Includes index. 1. Hyperfine structure. I. Title.
II. Series. QCl.S797 vol. 96 [QC173.4H95] 539s [539'.14] 82-10808

ISBN 978-3-662-15778-7   ISBN 978-3-540-39477-8 (eBook)
DOI 10.1007/978-3-540-39477-8

© by Springer-Verlag Berlin Heidelberg 1982
Originally published by Springer-Verlag Berlin Heidelberg New York in 1982.
Softcover reprint of the hardcover 1st edition 1982

2153/3130 — 543210

Preface

Recent advances in experimental techniques have opened up new fields of applications for the atomic beam magnetic resonance method and provided a broad spectrum of new information on the hyperfine structure of free atoms. This is particularly true for the transition elements with unfilled 4d or 5d electron shells. Their study using atomic beam magnetic resonance was made possible by the development of a universal method of producing intense atomic beams of these highly refractory elements.

The hyperfine structure of the 4d- and 5d-shell atoms is of particular interest, both from the standpoint of atomic physics, since most of these elements have many metastable atomic states, which are sufficiently populated at the evaporation temperature to permit atomic beam magnetic resonance studies, and from the standpoint of nuclear physics, since especially the 5d elements lie in a deformed region of the nuclear chart. In this book recent experimental results on the hyperfine structure of metastable states of 4d and 5d elements are described. At the same time, the modern theory of hyperfine structure based on the effective operator approach is reviewed. From an analysis of the data with respect to this formalism a considerable amount of new information is obtained concerning the nuclear ground-state properties as well as the atomic structure of the transition elements. Special emphasis is laid on discussing in detail the influences of relativistic and configuration interaction effects on the hyperfine interaction.

The aim of this monograph is to demonstrate — on the basis of a detailed analysis of hyperfine structure measurements in 4d- and 5d-shell atoms — that atomic beam magnetic resonance is a powerful technique for getting valuable information on both nuclear and atomic structure.

Bonn, July 1982   *S. Büttgenbach*

Contents

1. Introduction

In recent years investigations of hyperfine structure (hfs) and isotope shift (IS) in atomic spectra have provided a broad spectrum of new information concerning the electronic structure of free atoms and the properties of nuclei. This progress is connected with the application of new experimental techniques as well as with improvements in the theory of atomic fine and hyperfine structure.

The hyperfine interaction between the nucleus and the electrons of an atom can be represented, in general form, by an expansion in multipoles of order k,

$$H_{hfs} = \sum_k T^k(n) \cdot T^k(e)$$

where $T^k(n)$ and $T^k(e)$ are spherical tensor operators of rank k in the nuclear and electronic space, respectively.

The spherically symmetrical k=0 term is just the Coulomb interaction between the electrons and a point nucleus. Since the nucleus has a finite extension this term requires modification for electrons with a finite charge density within the nuclear volume. The energy shift of the fine structure levels due to this modification may be different for different isotopes of the same element because of the variation in the nuclear charge distribution with neutron number. This part of the IS, the so-called field (or volume) shift, is proportional to the change of the mean-square nuclear charge radius. The second contribution to the experimentally observed IS is the mass shift which originates from the recoil of the electron motion on the nucleus. It contains a trivial correction due to the change of the reduced electron mass (normal mass shift) and a contribution depending on correlations in the motion of the electrons (specific mass shift).

The k=1 term accounts for the interaction of the nuclear magnetic dipole moment with the magnetic field created at the nucleus by the atomic electrons. The k=2 term accounts for the interaction of the nuclear electric quadrupole moment with the inhomogeneous electric field produced by the electrons. The magnetic octupole (k=3) and electric hexadecapole (k=4) terms are nearly always much smaller than the

magnetic dipole and electric quadrupole terms, and so far these effects have been
observed in some favorable cases only.

Since hfs and IS depend on properties both of the nucleus and of the electrons
their investigation can give information on nuclear properties (spins, moments,
charge radii) as well as on atomic structure. In order to evaluate nuclear quantities from hfs and IS measurements the electronic part of the interaction has either
to be calculated by means of semiempirical or ab initio methods, or to be determined
from an independent measurement of the nuclear quantity for at least one isotope or
isotope pair of the same element. Magnetic moments are available for most stable
isotopes /1/ from atomic beam magnetic resonance (ABMR) or from nuclear magnetic
resonance (NMR). In contrast, most quadrupole moments are based on theoretical calculations of the average electric field gradient at the nucleus. However, nowadays
an increasing number of quadrupole moments are being determined nearly model independently from the hfs of mesonic atoms /2/. In order to extract the change of
the mean-square charge radius from IS measurements the optical data may be cali-
brated with respect to those obtained by measurements of muonic /3/ or electronic
X-ray shifts /4/ or by electron scattering /5/.

On the other hand, independent measurements of the hfs and the nuclear moments
allow a detailed study of the electronic structure of free atoms. The experimental
results may be compared to theoretical calculations, thus providing a rigorous test
of atomic hfs theory. The main difficulties in analyzing experimental hfs data arise
from the influence of relativistic and configuration interaction (CI) effects on the
hyperfine interaction. In order to take into account these effects the effective
operator formalism is frequently used. In 1965 it was found by HARVEY /6/ in his
analysis of the magnetic dipole hfs of oxygen and fluorine that a considerably improved fit to the experimental data could be obtained by use of a three-parameter
Hamiltonian. In order to make the inclusion of relativistic effects in the hfs of
many-electron atoms more convenient, SANDARS and BECK /7/ in 1965 introduced an
effective Hamiltonian that, when interacting between nonrelativistic atomic states,
produces the same result as the true Hamiltonian between relativistic states. This
operator contains three effective radial parameters for each open electron shell
and for each order k of the hyperfine interaction. It has been shown by JUDD /8,9/
and SANDARS /10/ that CI effects can be included in an effective Hamiltonian of
the same form. Therefore, in a modern analysis of hfs data the effective radial
parameters are treated as free parameters which are fitted to the experimental results in order to take into account relativistic and CI effects.

This parametric method can be applied if the hfs has been measured in a sufficiently large number of fine structure levels. For example, the hfs has to be known
in at least three levels for an electron configuration of type nl^N ($l \neq 0$). In 1974

LINDGREN and ROSEN /11/ analyzed a large number of experimental hfs data mainly in atomic ground configurations using the effective operator formalism. Their comparison of the experimental radial parameters with theoretical results obtained from relativistic Hartree-Fock calculations has demonstrated that there is appreciable influence of CI on the hyperfine interaction. However, the accurate ab initio calculation of the radial parameters is still a challenging problem. So far theoretical studies of CI using the multiconfiguration Hartree-Fock method or many-body perturbation theory have been performed for simple atoms only /12-14/.

Experimental investigations of the hfs require high spectroscopy resolution because the energy splitting of the fine structure levels caused by the hyperfine interaction is in general very small. The accuracy of classical optical spectroscopy by means of high-resolution interferometers is limited by the Doppler width of the spectral lines. A remarkable advance in experimental technique over that of optical spectroscopy was the development of the ABMR method by RABI and co-workers /15,16/. Using this method, precise values have been obtained for the hfs constants in ground and in low-lying metastable states of free atoms because the Doppler width is generally quite negligible in radio-frequency (rf) spectroscopy. In addition, a number of experimental techniques have been developed to enhance the population of metastable states and to increase the detection efficiency for metastable atoms, thus allowing ABMR investigations of high-lying metastable levels as well. In particular, the use of tunable dye lasers has led to a variety of modifications of the classical ABMR method opening new fields of application of radio-frequency spectroscopy.

Experimental details and results of ABMR have been described in several review articles /17-22/. A large number of stable and long-lived radioactive isotopes have been investigated yielding nuclear spins, magnetic dipole and electric quadrupole moments as well as electronic g_J factors and hfs interaction constants of metastable fine structure states /23,24/. Moreover, during recent years considerable progress has been made in the measurement of spins and moments of short-lived nuclei far from stability. For these measurements an ABMR apparatus has been connected on-line to the isotope separator ISOLDE at CERN /25/. Extensive investigations of short-lived excited atomic levels have been performed mainly by the optical double-resonance method, level crossing experiments, and optical pumping techniques /26-28/.

An ideal field for a detailed study of the influence of relativistic and CI effects in the hfs are the transition elements with an unfilled 3d, 4d, or 5d electron shell. These elements have many metastable states belonging to the configurations nd^N, $nd^{N-1}(n+1)s$, and $nd^{N-2}(n+1)s^2$. Therefore, in d-shell atoms the hfs can be measured in a number of different states providing a rigorous test of fine details of hfs-theory. First extensive measurements of the hfs in many metastable states of the same atom have been performed for the 3d-shell atoms /29,30/. There are also detailed theoretical predictions for the influence of CI on the effective radial parameters in the $3d^{N-2}4s^2$ and $3d^{N-1}4s$ configurations /31,32/. For the case of 4d- and 5d-shell atoms a theoretical hfs analysis is much more complicated because of the importance of relativistic effects and of strong spin-orbit and CI mixing in these atoms. Nevertheless, precise measurements of the hfs in metastable atomic states of 4d- and 5d-shell atoms are of great interest for the following reasons.

1) Such measurements form the basis for an improved understanding of atomic structure.
2) Atomic wavefunctions are usually obtained from a study of fine structure. Use of the fine-structure wavefunctions to calculate hfs effects is one of the most rigorous tests which can be applied to the wavefunctions. Thus, the analysis of precise hfs data is a sensitive indicator of the accuracy of atomic wavefunctions.
3) From the hfs measured in many atomic states a more reliable value of the nuclear electric quadrupole moment can be evaluated than from the hfs of the ground state alone.
4) From the hfs at high external magnetic field the nuclear magnetic dipole moment can be determined with high precision. In order to overcome the difficulties arising from hyperfine and Zeeman interactions with nearby atomic states such high-field measurements of the magnetic dipole moment have to be performed in several metastable atomic states.

Until recently, the 4d- and 5d-shell atoms, with very few exceptions, could not be studied by the ABMR method because of their high evaporation temperature. Since the development of a universal method of producing intense atomic beams of these refractory elements /33/, hfs measurements have been performed on most of them. These experiments are the subject of this review.

2. Theoretical Considerations

2.1 Traditional Hamiltonian for the Hyperfine Structure of Free Atoms

The hfs of the atomic energy levels is caused by the interaction between the orbital electrons and the electromagnetic moments of the nucleus. If we assume that the nuclear wavefunctions have well-defined parity, only magnetic moments of odd order and electric moments of even order will be non-zero. In addition, the magnitude of the hyperfine interaction energies decreases rapidly with order. Thus, the observable terms of the hyperfine interaction are the magnetic dipole term, the electric quadrupole term, and very small magnetic octupole and electric hexadecapole terms; the ratio of the magnetic octupole and magnetic dipole terms as well as the ratio of the electric hexadecapole and electric quadrupole terms are of the order of 10^{-5} /17/. The extremely small higher order terms can be neglected.

The hyperfine interaction can be conveniently represented by an expansion in multipoles of order k /34/,

$$H_{hfs} = \sum_{k \geq 1} T^k(n) \cdot T^k(e) \quad . \tag{1}$$

Here $T^k(n)$ and $T^k(e)$ are spherical tensor operators of rank k in the nuclear and electronic space, respectively, and each term in the expansion is the scalar product of these two tensors.

The electronic angular momentum $\underline{J}$ and the nuclear spin $\underline{I}$ couple to form distinct hfs states characterized by the total angular momentum quantum number F corresponding to the angular momentum operator $\underline{F} = \underline{I} + \underline{J}$. In the case of a point nucleus the first-order hfs energy displacement W_F from the unperturbed fine-structure level is

$$
\begin{aligned}
W_F &= \langle J\,I\,F | H_{hfs} | J\,I\,F \rangle \\
&= \sum_{k \geq 1} (-1)^{J+I+F} \begin{Bmatrix} J & I & F \\ I & J & k \end{Bmatrix} \langle I \| T^k(n) \| I \rangle \; \langle J \| T^k(e) \| J \rangle \quad .
\end{aligned} \tag{2}
$$

Strictly speaking, F is the only good quantum number when the hyperfine interaction is taken into account. On the other hand, because of the large energy differences between nuclear states and small hyperfine energy differences, we can make the approximation that I remains a good quantum number even when hyperfine interactions are considered. However, hyperfine interactions between the state under consideration and other nearby atomic states may perturbe the first-order energy levels of (2). If the hyperfine energies are assumed small compared to the fine-structure splitting, the breakdown of J as a good quantum number can be treated in second-order perturbation theory (Sect.2.7).

From the 6-j symbol in (2) it can be seen that the expansion (1) terminates with the term $k = \min(2I, 2J) = k_{max}$. We now define hyperfine interaction constants $A_k(J)$ /34/

$$A_k(J) = <I\,I\,|T_0^k(n)|I\,I> \; <J\,J\,|T_0^k(e)|J\,J> \quad, \tag{3}$$

and find for the hfs term energy

$$W_F = \sum_{k=1}^{k_{max}} (-1)^{I+J+F} \begin{Bmatrix} J & I & F \\ I & J & k \end{Bmatrix} \frac{A_k(J)}{\begin{pmatrix} J & k & J \\ -J & 0 & J \end{pmatrix}\begin{pmatrix} I & k & I \\ -I & 0 & I \end{pmatrix}} \quad. \tag{4}$$

The A's are related to the usual hfs interaction constants as follows:

$$\begin{aligned} A_1(J) &= h\,I\,J\,A\,(J) & A_2(J) &= (1/4)\,h\,B\,(J) \\ A_3(J) &= h\,C\,(J) & A_4(J) &= h\,D\,(J) \end{aligned} \tag{5}$$

in which h is Planck's constant. By explicit evaluation of the 3-j and 6-j symbols /35/, we get

$$W_F = h\,A(J)\,\frac{K}{2} + h\,B(J)\,\frac{(3/4)K(K+1)-I(I+1)J(J+1)}{2I(2I-1)J(2J-1)} \tag{6}$$

where $K = F(F+1)-I(I+1)-J(J+1)$. This expression takes into account only magnetic dipole and electric quadrupole interaction. Corresponding expressions for the magnetic octupole and electric hexadecapole interaction have been given by SCHWARZ /34/, for example.

2.2 Effective Operators

For a detailed study of the hyperfine interaction the effective operator formalism has been found to be extremely useful. The basic idea of the effective operator approach is to describe the electronic part of the hyperfine Hamiltonian by effective operators $T_{eff}^k(e)$, which act within a certain subspace (model space), normally a single configuration. In the case of several strongly interacting configurations it is also possible to choose a model space consisting of several configurations. The effect of admixtures from states outside the model space is thus a modification of the effective operator rather than a change of the wavefunction.

SANDARS and BECK /7/ have shown that, in addition to CI, relativistic effects on the hyperfine interaction can also be included in effective operators which interact between nonrelativistic LS coupled atomic states thus avoiding the difficulties involved in the use of relativistic jj coupled states.

Since the uncertainties in the values of the magnetic octupole and electric hexadecapole constants C and D evaluated from the presently available hfs data of 4d- and 5d-shell atoms are as large as the values themselves or even larger, we limit ourselves to a detailed discussion of the magnetic dipole and electric quadrupole interaction. The effective Hamiltonians for these interactions may be written as /11/

$$H_{dip}^{eff} = h \frac{T^1(n)}{\mu_I/I} \sum_i \left(a_{nl}^{01} \underline{l}_i - [10]^{1/2} a_{nl}^{12}(\underline{s}_i c_i^2)^1 + a_{nl}^{10} \underline{s}_i \right) \tag{7a}$$

$$H_Q^{eff} = h \frac{T^2(n)}{eQ} \sum_i \left(-b_{nl}^{02} c_i^2 + [3/10]^{1/2} b_{nl}^{13} U_i^{(13)2} + [3/10]^{1/2} b_{nl}^{11} U_i^{(11)2} \right) \tag{7b}$$

where the summation is limited to the open shells of the model space. In the case of an unpaired s electron only the term $a_{ns}^{10} \underline{s}_i$ contributes to the hfs energy. In the above expressions $\underline{s}_i$ and $\underline{l}_i$ are the spin and orbital angular momentum operators of the ith electron, c^2 is a second-rank tensor operator proportional to the spherical harmonic Y^2, and the $U^{(k_s k_1)k}$ are double tensor operators with rank k_s, k_1, and k in spin, orbital, and combined space, respectively /7/. The nuclear tensor operators $T^1(n)$ and $T^2(n)$ are defined in terms of the nuclear magnetic dipole moment μ_I and the electric quadrupole moment Q,

$$<I \ I \ |T_0^1(n)| \ I \ I> = \mu_I \tag{8a}$$

$$<I \ I \ |T_0^2(n)| \ I \ I> = (1/2)eQ \quad . \tag{8b}$$

The $a^{k_s k_1}$ and $b^{k_s k_1}$ are effective radial parameters which are treated as adjustable quantities to be fitted to the experimental data in order to take into account relativistic and CI effects. The radial parameters are related to the nuclear moments μ_I and Q, respectively, and to effective expectation values of $<r^{-3}>$ as follows:

$$a_{nl}^{k_s k_1} = \frac{2\mu_B}{h} \frac{\mu_I}{I} <r^{-3}>_{nl}^{k_s k_1} \qquad 1 > 0 \tag{9a}$$

$$a_{ns}^{10} = \frac{2\mu_B}{h} \frac{\mu_I}{I} <r^{-3}>_{ns}^{10} \qquad 1 = 0 \tag{9b}$$

$$b_{nl}^{k_s k_1} = \frac{e^2}{h} Q <r^{-3}>_{nl}^{k_s k_1} \qquad 1 > 0 \quad . \tag{9c}$$

The changes in the $<r^{-3}>$ values brought about by relativistic and CI effects will be discussed in detail in the following two sections.

2.3 Relativistic Effects

The Hamiltonian most commonly used to describe the interaction between relativistic electrons and a nucleus of charge Ze is given by

$$H^{rel} = \sum_i \left(c\underline{\alpha}_i \cdot \underline{p}_i + \beta_i mc^2 - \frac{Ze^2}{r_i} \right) + \sum_{i>j} \frac{e^2}{r_{ij}} + H_B \tag{10}$$

where H_B is the Breit interaction /36/

$$H_B = \sum_{i>j} \left\{ -\frac{e^2}{r_{ij}} \underline{\alpha}_i \cdot \underline{\alpha}_j + \frac{e^2}{2r_{ij}} \left(\underline{\alpha}_i \cdot \underline{\alpha}_j - \frac{(\underline{\alpha}_i \cdot \underline{r}_{ij})(\underline{\alpha}_j \cdot \underline{r}_{ij})}{r_{ij}^2} \right) \right\} \tag{11}$$

and $\underline{r}_{ij} = \underline{r}_i - \underline{r}_j$ and $r_{ij} = |\underline{r}_{ij}|$; $\underline{\alpha}$ and β are 4×4 matrices related to the unit matrix I and the Pauli spin matrices $\underline{\sigma}$ by

$$\underline{\alpha} = \begin{pmatrix} 0 & \underline{\sigma} \\ \underline{\sigma} & 0 \end{pmatrix} \qquad \beta = \begin{pmatrix} I & 0 \\ 0 & -I \end{pmatrix} \quad . \tag{12}$$

The first term in (11) corresponds to the magnetostatic interaction between the electrons; the second term represents the retardation effect of the electromagnetic interaction. Although the interaction (11) is only of order $(Z\alpha)^2$ relative to the Coulomb interaction, it contains all of the interactions which are, at present time, necessary in order to interpret the electronic structure of complex atoms. For a detailed discussion of the Breit interaction the reader is referred to BETHE and SALPETER /37/ and AKHIEZER and BERESTETSKII /38/.

In order to solve the eigenvalue problem for the Hamiltonian H^{rel} of (10), we start from the central field approximation; that is, each electron is assumed to move in a spherically symmetric field which is produced by the nuclear field and the spherically averaged fields of all other electrons. Thus, we write

$$H^{rel} = H_C^{rel} + H_P^{rel} + H_B \tag{13}$$

where

$$H_C^{rel} = \sum_i \left[c\underline{\alpha}_i \cdot \underline{p}_i + \beta_i mc^2 - \frac{Ze^2}{r_i} + U(r_i) \right] = \sum_i H_D(i)$$

and

$$H_P^{rel} = \sum_{i>j} \frac{e^2}{r_{ij}} - \sum_i U(r_i) \quad .$$

The central field potential $U(r)$ is chosen so as to make H_P^{rel} small. The Hamiltonian H_C^{rel} is simply a sum of single-particle Dirac Hamiltonians for an electron in a central field. The eigenfunctions of H_C^{rel} can be expressed as antisymmetric products of single-particle wavefunctions of the form

$$|\psi) = \begin{pmatrix} \frac{1}{r} F_{nlj}(r) \; |l \; j \; m_j> \\ \frac{i}{r} G_{nlj}(r) \; |\bar{l} \; j \; m_j> \end{pmatrix} \tag{14}$$

where $\bar{l} = 2j-1$. Then H_P^{rel} and H_B are treated as perturbations /39/.

The hfs Hamiltonian

$$H_{hfs}^{rel} = \sum_i \left[\frac{Ze^2}{r_i} - eV(r_i) + e\underline{\alpha}_i \cdot \underline{A}(r_i) \right] \tag{15}$$

in which $V(r)$ and $\underline{A}(r)$ are, respectively, the scalar and vector potentials of the nucleus is treated as perturbation, too. However, as pointed out in the preceding section, this is done using an effective operator H_{hfs}^{eff} of the form (7) whose matrix elements between nonrelativistic SL coupled states $|\psi>$ are the same as those of the relativistic Hamiltonian H_{hfs}^{rel} between the relativistic states $|\psi)$,

$$(\psi | H_{hfs}^{rel} | \psi) = <\psi | H_{hfs}^{eff} | \psi> \quad . \tag{16}$$

SANDARS and BECK /7/ derived explicit expressions for the effective radial parameters $a^{k_s k_l}$ and $b^{k_s k_l}$ in terms of integrals of relativistic radial wavefunctions, for $l \neq 0$:

$$\langle r^{-3}\rangle^{01}_{nl} = -\frac{e}{\mu_B}\frac{1}{(2l+1)^2}\left[-(l+1)P_{++} + P_{+-} + lP_{--}\right] \tag{17a}$$

$$\langle r^{-3}\rangle^{12}_{nl} = -\frac{e}{\mu_B}\frac{1}{3}\frac{1}{(2l+1)^2}\left[2(l+1)(2l-1)P_{++} - (2l-1)(2l+3)P_{+-} + 2l(2l+3)P_{--}\right] \tag{17b}$$

$$\langle r^{-3}\rangle^{10}_{nl} = \frac{e}{\mu_B}\frac{2}{3}\frac{1}{(2l+1)^2}\left[(l+1)^2 P_{++} + 2l(l+1)P_{+-} + l^2 P_{--}\right] \tag{17c}$$

$$\langle r^{-3}\rangle^{02}_{nl} = \frac{1}{(2l+1)^2}\left[(l+2)(2l-1)T_{++} + 6T_{+-} + (2l+3)(l-1)T_{--}\right] \tag{17d}$$

$$\langle r^{-3}\rangle^{13}_{nl} = -\frac{2}{(2l+1)}\left(\frac{2l(l+2)(l-1)l(l+1)}{5(2l+3)(2l+1)(2l-1)}\right)^{1/2}\left[(2l-1)T_{++} + 4T_{+-} - (2l+3)T_{--}\right] \tag{17e}$$

$$\langle r^{-3}\rangle^{11}_{nl} = -\frac{2}{(2l+1)}\left(\frac{6l(l+1)}{5(2l+1)}\right)^{1/2}\left[-(l+2)T_{++} + 3T_{+-} + (l-1)T_{--}\right] \tag{17f}$$

for l = 0:

$$\langle r^{-3}\rangle^{01}_{ns} = \langle r^{-3}\rangle^{12}_{ns} = 0 \tag{17g}$$

$$\langle r^{-3}\rangle^{10}_{ns} = \frac{e}{\mu_B}\frac{2}{3}P_{++} \quad . \tag{17h}$$

The signs + and − refer to the case j = l+1/2 and j = l−1/2, respectively, and the radial integrals $P_{jj'}$ and $T_{jj'}$ are defined by

$$P_{jj'} = \int \frac{F_{nlj}\,G_{nlj'} + G_{nlj}\,F_{nlj'}}{r^2}\,dr \tag{18a}$$

$$T_{jj'} = \int \frac{F_{nlj}\,F_{nlj'} + G_{nlj}\,G_{nlj'}}{r^3}\,dr \quad . \tag{18b}$$

In the nonrelativistic limit we find that

$$\langle r^{-3}\rangle^{01}_{nl} = \langle r^{-3}\rangle^{12}_{nl} = \langle r^{-3}\rangle^{02}_{nl} = \langle r^{-3}\rangle_{nl} = \int \frac{R^2_{nl}}{r^3}\,dr \quad ,$$

$$\langle r^{-3}\rangle^{10}_{nl} = \langle r^{-3}\rangle^{13}_{nl} = \langle r^{-3}\rangle^{11}_{nl} = 0, \quad \text{and}$$

$$\langle r^{-3}\rangle_{ns}^{10} = \langle r^{-3}\rangle_{ns} = \frac{2}{3}\left|\frac{R_{ns}}{r}\right|_{r=0}^{2} = \frac{8\pi}{3}\left|\psi(0)\right|_{ns}^{2} \tag{19}$$

where R_{nl} is the nonrelativistic limit of F_{nlj}, which is obtained by solving the nonrelativistic radial Schrödinger equation, and $|\psi(0)|_{ns}^{2}$ is the density of the s electron at the nucleus.

Thus, relativistic effects in the hyperfine interaction produce nonvanishing radial integrals $\langle r^{-3}\rangle^{10}$, $\langle r^{-3}\rangle^{13}$, and $\langle r^{-3}\rangle^{11}$ and shift $\langle r^{-3}\rangle^{01}$, $\langle r^{-3}\rangle^{12}$, and $\langle r^{-3}\rangle^{02}$ from their nonrelativistic value $\langle r^{-3}\rangle$. For unpaired s electrons the quantity $\langle r^{-3}\rangle_{ns}^{10}$ is changed from the nonrelativistic value $\langle r^{-3}\rangle_{ns}$. It is convenient to introduce relativistic correction factors (RCF) for the effective expectation values of $\langle r^{-3}\rangle$. For the magnetic dipole and electric quadrupole interaction the RCF are defined as /11/

$$F_{nl}^{k_s k_l} = \frac{\langle r^{-3}\rangle_{nl}^{k_s k_l}}{\langle r^{-3}\rangle_{nl}} \qquad \text{for } k_s k_l = 01,\ 12,\ 10 \tag{20a}$$

$$R_{nl}^{k_s k_l} = \frac{\langle r^{-3}\rangle_{nl}^{k_s k_l}}{\langle r^{-3}\rangle_{nl}} \qquad \text{for } k_s k_l = 02,\ 13,\ 11 \quad . \tag{20b}$$

In practice, one calculates the quantities $\langle r^{-3}\rangle^{k_s k_l}$ with radial wavefunctions obtained using some type of self-consistent-field (SCF) calculation. The status of relativistic atomic structure calculations has been reviewed by GRANT /40/, LINDGREN and ROSEN /11/, and DESCLAUX /41/, and we shall give only an outline of the method here. Starting from a suitably chosen central potential U(r), one finds zero-order wavefunctions by solving the eigenvalue equation for the Hamiltonian H_c^{rel} in (13). The relativistic Hartree-Fock equations are obtained using the variational principle applied to the expectation value of the total Hamiltonian with respect to the zero-order wavefunction and are solved in a self-consistent way.

In the unrestricted Hartree-Fock (UHF) scheme the single-electron orbitals are allowed to vary without any restrictions besides the orthonormality condition. In the restricted Hartree-Fock (RHF) scheme the orbitals are of the form (14) and only the radial part is varied. The radial equations for F_{nlj} and G_{nlj} are obtained by applying the variational principle after working out the angular parts in the evaluation of the expectation value of the total Hamiltonian. Because the spin-orbit interaction is explicitly taken into account in the one-electron Dirac Hamiltonian the total angular momentum J of the electrons is a good quantum number, but not the orbital and spin angular momenta L and S. As a result, a given nonrelativistic single LS configuration will be split into various jj subconfigurations. Since LS coupling dominates for a large number of atoms it is necessary to perform calculations with more than a single jj configuration even at the Hartree-Fock level. For this purpose the multiconfiguration Dirac-Fock (MCDF) method which can handle a priori the inter-mediate coupling scheme has been developed /42,43/.

In principle, MCDF calculations lead to different solutions for different states of a configuration. However, for many purposes it is convenient to use the same ra-dial functions for all states within a configuration. In such cases the complexity of the problem can be reduced by calculating the wavefunctions by minimizing the average energy of a configuration instead of carrying out a separate calculation for each level. Various averaging techniques have been proposed /11,43,44/. In fact, for a single open shell the term dependence is usually fairly small /14/. However, when two open shells are present, the term dependence may be significant. Examples are the ^{3}P and ^{1}P terms of the lowest ns np configurations of the group II elements. The exchange interaction between the electrons ns and np has the effect of contract-ing the np function for the ^{3}P term and expanding it for ^{1}P /14,45,46/. In such cases the calculations should be performed for the proper term if high accuracy is needed. Without discussing further details of the MCDF method it should just be mentioned that this method is also able to handle correlation effects (Sect.2.4).

Existing relativistic calculations of atomic structure take into account the electron-electron interaction in a two-step process: while the Coulomb interaction is introduced in the Hamiltonian used to determine the wavefunctions, which is thus $H_{HF}^{rel} = H_C^{rel} + H_P^{rel}$, the Breit interaction is treated in first-order perturbation the-ory to partially correct for relativistic effects in the electron-electron inter-action. To include the Breit interaction in the self-consistent procedure has been claimed to be incorrect /45,47/. Recently, BUCHMÜLLER /48/ and SUCHER /49/ contra-dicted this rule. They considered variational approximations to the bound-state problem in many-electron atoms and derived within the framework of quantum electro-dynamics SCF equations which include the Breit interaction or part of it. These approaches could be of particular importance in the case of heavy atoms where the Coulomb repulsion of the electrons, their magnetic interaction, and retardation effects of the electromagnetic interaction are all of the same order of magnitude. However, the question which interactions the Hamiltonian should comprise can only be answered by a numerical analysis.

2.4 Effects of Configuration Interaction

In order to take into account configuration mixing in a systematic way, atomic structure calculations are nowadays frequently based on the linked diagram expansion developed by BRUECKNER /50/ and GOLDSTONE /51/, which was first applied to atomic problems by KELLY /52/. This method, which leads to a many-body perturbation theory (MBPT) procedure, can be combined in a very nice way with the effective-operator formalism, as has, for instance, been shown by SANDARS /10,53/ and JUDD /54,55/.

Because of the complexity of the problem one usually starts from the simple nonrelativistic Hamiltonian

$$H_S = \sum_i \left(\frac{p_i^2}{2m} - \frac{Ze^2}{r_i} \right) + \sum_{i>j} \frac{e^2}{r_{ij}} \quad . \tag{21}$$

The purpose of the linked diagram perturbation technique is then to yield an approximate solution of the Schrödinger equation

$$H_S \, |\psi^a\rangle = E^a \, |\psi^a\rangle \quad . \tag{22}$$

In a first step the Hamiltonian H_S is split into a model Hamiltonian H_C and a perturbation H_P,

$$H_S = H_C + H_P \quad . \tag{23}$$

It is convenient to choose H_C to be a central field Hamiltonian of the form

$$H_C = \sum_i \left(\frac{p_i^2}{2m} - \frac{Ze^2}{r_i} + U(r_i) \right) \quad . \tag{24}$$

The eigenfunctions of H_C are Slater determinants constructed out of single-particle wavefunctions of the form

$$|\phi\rangle = \frac{1}{r} R_{nl}(r) \, |l \, m_s \, m_l\rangle \quad . \tag{25}$$

The perturbation is then the noncentral part of the Coulomb interaction between the electrons,

$$H_P = \sum_{i>j} \frac{e^2}{r_{ij}} - \sum_i U(r_i) \quad . \tag{26}$$

The eigenfunctions of H_C are used to partition the Hilbert space into the model space and the orthogonal space. The model space consists of all states of one or several configurations. For example, if one is interested in states belonging to the configuration $nd^{N-2}(n+1)s^2$, which is the ground configuration in many d-shell atoms, the model space may be chosen to consist of all states of this configuration. All other states belong to the orthogonal space.

The next step is to transform the full Hamiltonian H_S operating in the entire Hilbert space into an effective Hamiltonian H^{eff} operating only in the model space. States which originate from the model space, that is, states which would go into the model space if the perturbation is tuned off slowly, can then be described by this effective operator. If the model space has d dimensions there are, in general, d such states $|\psi^a\rangle$ (a = 1,2...d). The effective Hamiltonian satisfies the equation

$$H^{eff} |\psi_0^a\rangle = E^a |\psi_0^a\rangle \qquad (a = 1,2...d) \tag{27}$$

where the eigenfunctions $|\psi_0^a\rangle$ are the projections of the true wavefunctions $|\psi^a\rangle$ onto the model space. The effective operator reproduces all d eigenvalues of the full Hamiltonian. Thus, if we know the effective Hamiltonian with sufficient accuracy, we can obtain the exact energies of all states originating from the model space by solving a secular equation of finite dimensionality. The model space may be extended to include several strongly interacting configurations, for example the three configurations nd^N, $nd^{N-1}(n+1)s$, and $nd^{N-2}(n+1)s^2$ [1]. This makes it possible to treat their mixing exactly at the expense of solving a somewhat larger secular equation.

In the case of the hfs one is not primarily interested in the total energy of the system but rather in the small energy splitting due to the hyperfine interaction H_{hfs}. An effective hfs operator is obtained by replacing the perturbation H_p by $H_p + H_{hfs}$ and keeping in the perturbation expansion only terms linear in H_{hfs}. The energy shift caused by the hyperfine interaction is given by

$$E_{hfs}^a = \langle\psi_0^a| H_{hfs}^{eff} |\psi_0^a\rangle \tag{28}$$

where the functions $|\psi_0^a\rangle$ are eigenfunctions of H^{eff}.

---

[1] In the following we shall use the short-form notations d^N, $d^{N-1}s$, and $d^{N-2}s^2$ for the single configurations and the notation $(d+s)^N$ for all three configurations.

Then the linked diagram theorem states that /53/:

1) H_{hfs}^{eff} can be expanded in a perturbation series

$$H_{hfs}^{eff} = h_0^{eff} + h_1^{eff} + h_2^{eff} + \ldots \qquad (29)$$

such that h_n^{eff} is of order $H_{hfs} \cdot (H_p/\Delta E)^n$ where ΔE is an energy denominator of the order of single-particle energy differences between closed or open orbitals and excited orbitals.

2) Each term in the expansion (29) can be written as the sum of a constant, a one-body operator, a two-body operator, and so on up to an N-body operator where N is the number of electrons in open shells. Thus,

$$h_n^{eff} = f^n + \sum_i f^n(i) + \sum_{i \neq j} f^n(i,j) + \ldots \qquad . \qquad (30)$$

3) The matrix elements of these many-body operators can be represented graphically by means of Feynman diagrams. Interactions are represented by horizontal dotted lines and orbitals by directed vertical lines. We distinguish between orbitals belonging to closed, excited, and open shells in the model space (Fig.1a,b). The diagram representing an effective operator of the order n contains n interaction lines of which one and only one interaction line must be the hfs. One-body operators are represented by diagrams with one incoming and one outgoing line, two-body operators by diagrams with two incoming and two outgoing lines, etc. (Fig.1c). Only such diagrams contribute to the effective operators which have been constructed according to certain rules /10/. These are the so-called linked and backwards diagrams.

The proof of the linked diagram theorem is given by SANDARS /10/ and LINDGREN /56/, for example.

The zero-body operators have no angular dependence and give no hyperfine splitting. The most important parts of the effective hfs Hamiltonian are the one-body and two-body operators; n-body contributions to the hfs with $n \geq 3$ are usually neglected.

The first-order effective Hamiltonian is a one-body operator which is represented by the diagram in Fig.1d. Evaluation of the corresponding matrix elements yields the hfs constants in the central-field approximation.

The second-order effective Hamiltonian contains one- and two-body operators. Examples of second-order diagrams are given in Fig.1e-h. These diagrams have in the intermediate state a single excitation from closed or open shells. Three kinds of excitations are possible:

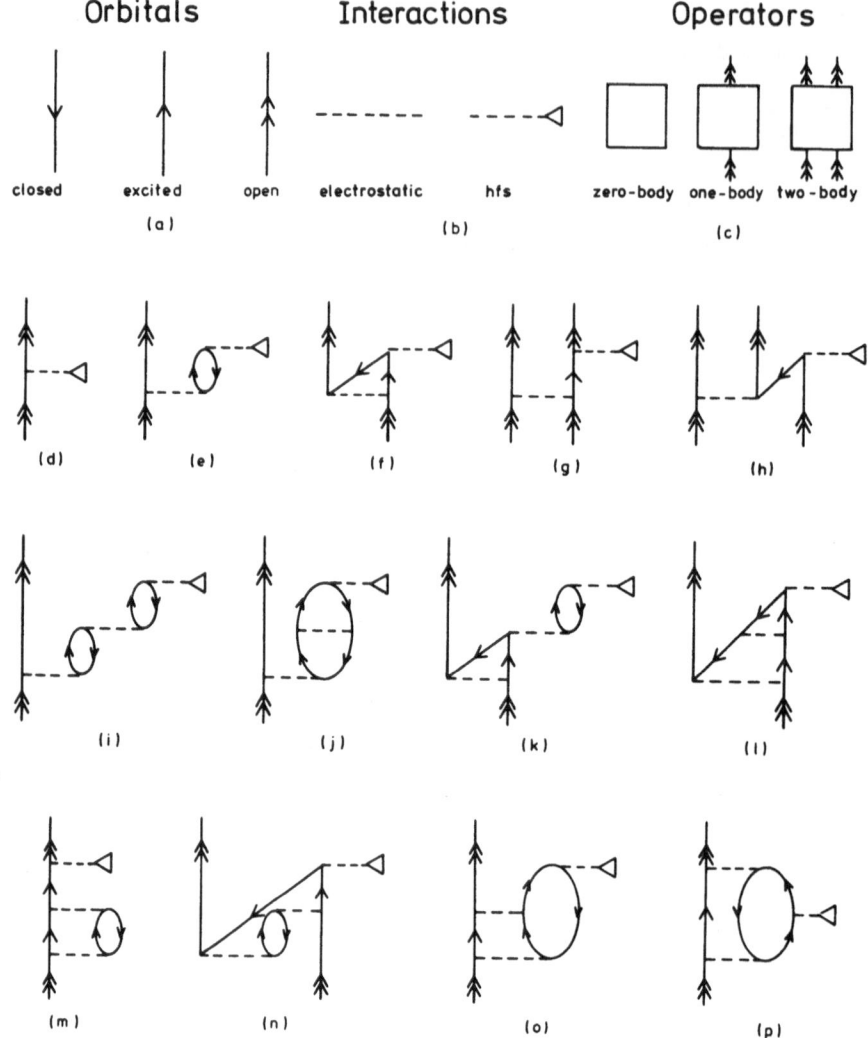

Fig.1a-p. Diagrammatic representation of effective hfs operators

1) Excitation of an electron of a closed shell to an excited shell.
2) Excitation of an electron of a closed shell to an open shell.
3) Excitation of an electron of an open shell to an excited shell.

Excitations of type 1) are described by one-body operators (Fig.1e,f), and excitations of type 3) by two-body operators (Fig.1g). Excitations of type 2) lead to one-body as well as two-body effects (Fig.1h).

The diagrams in Fig.1i-p are examples of third-order effective one-body diagrams. In this order we have for the first time diagrams which involve at least one double excitation. Effects which in each order are described by simultaneous excitations of several electrons are called correlation effects, whereas perturbations which are characterized by the fact that there is only one excitation at a time are referred to as polarization effects. Thus, diagrams (i-l) represent third-order polarization, and diagrams (m-p) third-order correlation effects. Since the second-order diagrams contain only single excitations, the second-order effect is entirely a polarization effect.

An analysis of the tensor structure of the one-body hfs operators shows that they have the same spin-angular structure as the Hamiltonians (7). Thus, the part of CI (polarization and correlation effects) that is represented by one-body operators can be incorporated in the Hamiltonians of (7) through a modification of the appropriate radial parameters (9). Furthermore, these modified radial parameters are the same for all states of a configuration. Distortions of inner closed s-shells, which can be described by single excitations of s-s type, lead to a modification of the contact term $<r^{-3}>^{10}$, an effect which is called spin polarization. Spin polarization can also be taken into account in the HF procedure if the radial functions associated with the two spin orientations are allowed to be different. This is the spin-unrestricted Hartree-Fock (SUHF) method, a simplified form of the general UHF method. For a detailed discussion of the various HF methods and their application to atomic structure calculations the reader is referred to FROESE-FISCHER /14/. Other types of polarization, which are represented by single excitations of the types p-p, d-d, s-d, p-f, etc., and correlation effects modify all radial parameters of the magnetic dipole interaction and the parameter $<r^{-3}>^{02}$ of the electric quadrupole interaction. The parameters $<r^{-3}>^{13}$ and $<r^{-3}>^{11}$ remain unchanged.

The second-order two-body operators, which describe the excitations of an electron of an open shell to an excited shell and the excitation of an electron of a closed shell to an open shell, have been studied in detail by ARMSTRONG /39/ and BAUCHE-ARNOULT /31,32/. Their influence can also be incorporated in the Hamiltonians (7) through a modification of the radial integrals $<r^{-3}>^{k_s k_l}$, where $k_s k_l = 01$, 12, 10, 02. However, these contributions are not constant for the states of a configuration but only for the terms of a SL multiplet. Therefore, if we take into account all CI effects that can be described by one-body effective operators as well as the second-order two-body effects, we can express the effective radial integrals in the following way:

$$<r^{-3}>^{01}_{nl} = F^{01}_{nl} <r^{-3}>_{nl} (1 + \Delta^{01}_{nl} + \Delta^{01}_{nl\text{-}SL}) \qquad (31a)$$

$$<r^{-3}>^{12}_{nl} = F^{12}_{nl} <r^{-3}>_{nl} (1 + \Delta^{12}_{nl} + \Delta^{12}_{nl\text{-}SL}) \qquad (31b)$$

$$\langle r^{-3}\rangle_{nl}^{10} = F_{nl}^{10} \langle r^{-3}\rangle_{nl} + \langle r^{-3}\rangle_{CI}^{10} + \langle r^{-3}\rangle_{CI-SL}^{10} \tag{31c}$$

$$\langle r^{-3}\rangle_{ns}^{10} = F_{ns}^{10} \langle r^{-3}\rangle_{ns} (1 + \Delta_{ns}^{10} + \Delta_{ns-SL}^{10}) \tag{31d}$$

$$\langle r^{-3}\rangle_{nl}^{02} = R_{nl}^{02} \langle r^{-3}\rangle_{nl} (1 + \Delta_{nl}^{02} + \Delta_{nl-SL}^{02}) \quad . \tag{31e}$$

Recently, the importance of SL-dependent CI has been demonstrated by JOHANN, DEMBCZYNSKI, and ERTMER /30,57,58/ in their analyses of the hfs of 3d-shell atoms. Since there is strong spin-orbit mixing in 4d- and 5d-shell atoms, in general, many SL basis states contribute to the eigenvectors in intermediate coupling. This makes the analysis with respect to SL-dependent effects in the hfs of these atoms very complicated. In addition, such an analysis requires precise knowledge of the hfs of more levels than have been investigated up to now with sufficient accuracy. Therefore, in the present analysis of the hfs of 4d- and 5d-shell atoms SL-dependent effects are neglected; that is, the expectation values $\langle r^{-3}\rangle^{k_s k_l}$ are assumed to be the same for all states of a configuration.

The modification of the electric quadrupole interaction parameter $\langle r^{-3}\rangle^{02}$ is often expressed as

$$\langle r^{-3}\rangle_{nl}^{02} = (1 - R_{nl}) \langle r^{-3}\rangle_{nl} \quad . \tag{32}$$

Angular excitations (n'l'-n''l'', l''≠l') contribute positively to R_{nl}, whereas radial excitations (n'l'-n''l') contribute negatively to R_{nl}. We speak of the former excitation as "shielding" the nuclear quadrupole moment (lessening the magnitude of the gradient of the electric field at the nucleus), and the latter as "anti-shielding" the quadrupole moment. The influence of polarization effects on the quadrupole interaction has been calculated by STERNHEIMER /59-63/ for many elements. However, as pointed out by LINDGREN /64/, correlation effects may also contribute appreciably to $\langle r^{-3}\rangle^{02}$. A compilation of "Sternheimer factors" R_{nl} is given in /39/.

If the model space consists of several configurations, for example, the configurations $(d+s)^N$, matrix elements of the hfs Hamiltonian between states belonging to different configurations, contribute to the total hfs splitting. These contributions are proportional to the radial integral $\langle nd |r^{-3}| (n+1)s\rangle$. Using HF wavefunctions very small values are found for this integral. Therefore, hyperfine interactions between the configurations are usually neglected. However, CI effects may strongly influence the off-diagonal hfs matrix elements as has been shown in a theoretical calculation by BAUCHE and BAUCHE-ARNOULT /65/ in the case of ScI and TiI. Empirically these contributions have been found to be significant in the case of ReI /66/; since the hfs of 15 levels belonging to the configurations $(5d+6s)^7$ had been measured with high precision, the off-diagonal radial parameters could be determined in a least-squares fit procedure.

Up to now, only very few numerical calculations of atomic hfs using the nonrelativistic MBPT procedure have been reported. The present state of the theory and its application in atomic physics can be seen from a recent review by LINDGREN and MORRISON /12/ and from the contributions to the 1979 Nobel symposium /67/. Accurate hfs calculations can, however, also be performed in a nonperturbative way, based on the variational principle. A summary of the various methods is given by HIBBERT /68/. One of the main variational methods is the nonrelativistic multiconfiguration Hartree-Fock (MCHF) method. In the MCHF approximation the wavefunction of a SL multiplet is a linear combination of single-configuration wavefunctions,

$$|\gamma SL> = \sum_i c_i \ |\gamma_i SL> \tag{33}$$

where the sum is, in principle, infinite, but for all practical calculations is finite. Then, in the SCF procedure not only the radial functions are determined variationally, but also the mixing coefficients c_i. Using the MCHF method, CI effects in the hfs of several atoms could be interpreted /14,46,65,68/.

The compromise which has been made in taking into account relativistic as well as CI effects on the hyperfine interaction is that these two effects are treated as "additive", that is, first the hfs is calculated using relativistic wavefunctions obtained in a central field, and then CI effects which are calculated using nonrelativistic wavefunctions are added. FENEUILLE and ARMSTRONG /69,70/ have considered possible corrections to this additive approach. They found corrections due to errors introduced through

1) the use of wavefunctions containing spin-orbit effects calculated in an approximate central field, and
2) the neglect of the Breit interaction in the relativistic central field.

The corrections, which are of order $(Z\alpha)^2$ smaller than the hfs itself, can be absorbed into the effective hfs Hamiltonian by modifying the effective radial integrals. As in the case of CI, there are contributions which are constant for all states of a configuration as well as SL-dependent contributions. These corrections also affect $<r^{-3}>^{13}$ and $<r^{-3}>^{11}$, and we have to add to (31a-e) the relations

$$<r^{-3}>_{nl}^{13} = R_{nl}^{13} <r^{-3}>_{nl} (1 + \Delta_{nl}^{13} + \Delta_{nl-SL}^{13}) \tag{31f}$$

$$<r^{-3}>_{nl}^{11} = R_{nl}^{11} <r^{-3}>_{nl} (1 + \Delta_{nl}^{11} + \Delta_{nl-SL}^{11}) \quad . \tag{31g}$$

It is clear that for a complete understanding of the hfs a relativistic MBPT treatment is necessary. A first step in this direction is the work of DAS, ANDRIESSEN, and co-workers. They introduced a relativistic formulation of the linked diagram perturbation analysis. Using this procedure they were able to explain for the first time hfs data in N, Mn, Rb, Eu, Gd, and Tl /13,71-73/.

2.5 Eigenvectors in Intermediate Coupling for Complex Atoms

According to (7,28) the hfs energy can be expressed in terms of the radial parameters $a^{k_s k_l}$ and $b^{k_s k_l}$ if eigenvectors $|\psi_0^a>$ of the effective fine-structure Hamiltonian H^{eff} are accurately known. In order to take partly into account relativistic effects in the fine-structure interaction, we construct an effective Hamiltonian that contains relativistic correction terms in addition to the Schrödinger Hamiltonian of (21). For this purpose we expand the relativistic Hamiltonian of (13) in powers of $Z\alpha$ and keep terms of the order $(Z\alpha)^2 E_C$, where E_C is the zero-order energy. Details of this procedure can be found in /70,74/. Second-order CI effects are taken into account in an effective way by adding an operator H_{CI} which will be discussed in detail below. Thus, we have

$$H^{eff} = H_C + H_P + H_R + H_{SO} + H_{SOO} + H_{OO} + H_{SS} + H_{SSC} + H_{CI} \tag{34}$$

where H_C is the central field Hamiltonian (24) and H_P is the noncentral part of the electrostatic interaction between the electrons (26). H_R contains a relativistic mass correction and relativistic corrections to the potential energy, the one- and two-body Darwin terms /75/. H_{SO} represents the usual central field spin-orbit interaction,

$$H_{SO} = \sum_i \xi(r_i) \, \underline{s}_i \cdot \underline{l}_i \tag{35}$$

where

$$\xi(r_i) = - \frac{\hbar^2}{2m^2 c^2} \frac{1}{r_i} \frac{d\varphi(r_i)}{dr_i} \quad , \tag{36}$$

with $\varphi(r_i)$ the potential felt by each electron. H_{SOO}, H_{OO}, H_{SS}, and H_{SSC} represent, respectively, the spin-other orbit, the orbit-orbit, the spin-spin, and the spin-spin contact interaction between the electrons.

In order to determine the eigenvectors $|\psi_0^a\rangle$ (a = 1, ..., d) one has to set up and to diagonalize the complete matrix of H^{eff}. The radial integrals which occur in the matrix elements are usually treated as free parameters and adjusted to give least-squares fits of the calculated levels to the observed ones. This also allows taking such relativistic and CI effects which are not included explicitly in the effective Hamiltonian of (34) partly into account. For setting up the energy matrix the SL coupling scheme is most widely used, since many tables and computer codes for the evaluation of matrix elements are based on this coupling scheme. In the case of the configurations $(d+s)^N$ [2] the seniority v /76/ is used as an additional quantum number in order to distinguish the different multiplets with the same quantum numbers S and L belonging to the same configuration. A detailed discussion of the classification schemes of states of mixed configurations $(l+l')^N$ using group-theoretical methods can be found in /77/. Since the real atomic states deviate from pure SL coupling, the configuration $\tau = d^N$, $d^{N-1}s$, or $d^{N-2}s^2$, the seniority v, and the spin and orbital angular momentum quantum numbers S and L are no longer good quantum numbers. However, the Hamiltonian H^{eff} commutates with $\underline{J}^2$, where $\underline{J} = \underline{S} + \underline{L}$ is the total electronic angular momentum. Thus, the construction of the matrix is simplified by the fact that it breaks up into a different, independent submatrix for each possible value of J.

In intermediate coupling a real atomic state can then be expressed as a linear combination of SL basis states of the same J,

$$|\psi_0^a\rangle = |\alpha J\rangle = \sum_{\tau vSL} \langle \tau vSLJ | \alpha J\rangle | \tau vSLJ\rangle \quad . \tag{37}$$

In this expression the real states are identified by their J value and the quantity α which, for example, may be the excitation energy of the state. In some cases one SL component of the eigenvector is large and all others are small. In this case the states may also be identified by the dominating SL component. The mixing coefficients $\langle \tau vSLJ | \alpha J\rangle$ are obtained from the unitary matrix which diagonalizes the energy matrix.

In the following we shall discuss the various contributions to the effective Hamiltonian in more detail.

---

[2] In the following we limit ourselves to configurations of this type. A generalization to more complex configurations including p and f electrons can easily be made.

1) The Hamiltonian H_C and the second term of H_p are purely radial terms, and their energy contributions have the same values for all levels belonging to a configuration. Therefore they do not affect the relative position of the levels of any one configuration. Furthermore, it can be shown /78/ that the contribution of those terms of the Coulomb interaction for which both electrons belong to closed shells or one electron is in a closed shell and one electron in an open shell is the same for all levels within a configuration. Thus, the summation in the first part of H_p can be limited to electrons belonging to open shells. This contribution is different for different states and thus causes the splitting of the SL terms of a configuration. The corresponding matrix elements are independent of J and diagonal in S and L, but not in v and τ. They can be expressed as linear combinations of the so-called Slater integrals

$$R^k(ab,cd) = e^2 \iint R_{n_a l_a}(r_1) R_{n_c l_c}(r_1) \frac{r_<^k}{r_>^{k+1}} R_{n_b l_b}(r_2) R_{n_d l_d}(r_2) \, dr_1 dr_2 \quad . \tag{38}$$

Using the abbreviations

$$R^k(ab,ab) = F^k(a,b) \quad \text{and} \quad R^k(ab,ba) = G^k(a,b) \tag{39}$$

and taking into account that k, l_a, and l_b must satisfy the triangular condition $|l_a - l_b| \le k \le l_a + l_b$ and that k must satisfy the condition $k + l_a + l_b =$ even /78/, we find the appropriate Slater integrals describing the Coulomb interaction within each of the three configurations $(d+s)^N$, namely,

for d^N and $d^{N-2}s^2$: $F^0(d,d)$, $F^2(d,d)$, and $F^4(d,d)$;

for $d^{N-1}s$: $F^0(d,d)$, $F^0(d,s)$, $F^2(d,d)$, $F^4(d,d)$, and $G^2(d,s)$.

Instead of the parameters F^0, whose angular coefficients have the same value for all levels of a configuration, one parameter E^0 is frequently used for each configuration representing the average energy of the configuration. Thus, E^0 comprises all contributions to the diagonal elements of the energy matrix which have the same value for all levels of the configuration. The electrostatic interaction between the configurations of the model space leads to additional radial parameters, namely,

for the interaction between d^N and $d^{N-1}s$: $H^2(d^N, d^{N-1}s) = R^2(dd,ds)$;

for the interaction between $d^{N-1}s$ and $d^{N-2}s^2$: $H^2(d^{N-1}s, d^{N-2}s^2) = R^2(dd,ds)$;

for the interaction between d^N and $d^{N-2}s^2$: $G^2(d^N, d^{N-2}s^2) = R^2(dd,ss)$.

The angular coefficients of the Slater parameters can be evaluated using tensor algebra techniques /76,79,80/. Explicit expressions as well as tabulations for certain types of configurations can be found in /76,79-83/, for example.

2) The next important term of H^{eff} is the spin-orbit interaction H_{SO} which removes the J degeneracy of the terms of a SL multiplet and which is represented by the spin-orbit parameters $\zeta(d^N)$, $\zeta(d^{N-1}s)$, and $\zeta(d^{N-2}s^2)$.

3) Considering these dominant terms of the Hamiltonian for heavy atoms, the electrostatic interaction H_P and the spin-orbit interaction H_{SO}, we define an effective operator H_{CI} which takes into account, to second order of perturbation theory, the interaction with configurations outside the model space, which are obtained from the $(d+s)^N$ configurations by excitation of one or two electrons. For each perturbing configuration the CI Hamiltonian decomposes into three parts, one purely electrostatic, a second which is purely spin-orbit, and a mixed term containing both electrostatic and spin-orbit interactions,

$$\Omega_1 = - \frac{1}{\Delta E} \sum_m H_P |m><m| H_P \tag{40a}$$

$$\Omega_2 = - \frac{1}{\Delta E} \sum_m H_{SO} |m><m| H_{SO} \tag{40b}$$

$$\Omega_3 = - \frac{1}{\Delta E} \sum_m (H_{SO} |m><m| H_P + H_P |m><m| H_{SO}) \tag{40c}$$

where the sum is over all states $|m>$ of the perturbing configuration. ΔE is the mean energy separation between the perturbing and the perturbed configuration. For the case of mixed configurations $(l+l')^N$ this effective Hamiltonian has been discussed in detail by FENEUILLE /84,85/ and SCHRIJVER and NOORMAN /86,87/.

For the configurations $(d+s)^N$ the effect of the electrostatic interactions Ω_1 is partly absorbed in the parameters E^0 and in the Slater integrals F^2, F^4, G^2, and H^2 if we use three different parameters $F^k(d^N)$, $F^k(d^{N-1}s)$, and $F^k(d^{N-2}s^2)$ for $k = 2,4$, two different parameters $H^2(d^N,d^{N-1}s)$ and $H^2(d^{N-1}s,d^{N-2}s^2)$, and two different parameters $G^2(d^{N-1}s)$ and $G^2(d^N,d^{N-2}s^2)$. However, the parameters F^k obey the rule /86/,

$$F^k(d^N) - F^k(d^{N-1}s) = F^k(d^{N-1}s) - F^k(d^{N-2}s^2) \quad \text{for } k = 2,4 \quad . \tag{41}$$

The effects not absorbed in the Slater integrals can be accounted for by introducing three two-body operators

$$\alpha \underline{L}^2 + \beta G(R_5) + \gamma G(R_6) \tag{42}$$

and several three-body operators. Here $\underline{L}$ is the orbital angular momentum operator, and $G(R_5)$ and $G(R_6)$ are Casimir operators for the groups R_5 and R_6, respectively /88/; α, β, and γ are adjustable radial parameters. The matrix elements of the operator (42) as well as the matrix elements of the effective three-body operators can be calculated using group theoretical methods /85,89/.

The effect for the purely spin-orbit terms Ω_2 can be fully absorbed in the first-order spin-orbit parameters.

The effect of the electrostatically correlated spin-orbit interactions Ω_3 can be expressed, after omitting terms proportional to the first-order spin-orbit interaction, by an effective Hamiltonian defined by PASTERNAK and GOLDSCHMIDT /90/,

$$H_{EL-SO} = -2 \sum_{k \text{ even}} Q^k \left[1(1+1)(21+1)/(2k+1) \right]^{1/2} \cdot \sum_{t \text{ odd}} (2t+1) \cdot$$

$$\begin{Bmatrix} 1 & k & t \\ 1 & 1 & 1 \end{Bmatrix} (U^k \cdot T^{(1t)k}) \tag{43}$$

where

$$Q^k = <1 \| C^k \| 1> \sum_{n'} R^k(n1,n1;n1,n'1) \cdot \zeta(n1,n'1)/\Delta E(n1,n'1) \tag{44}$$

with $R^k(n1,n1;n1,n'1)$ and $\zeta(n1,n'1)$, respectively, a Slater and a spin-orbit parameter; C^k, U^k, and $T^{(1t)k}$ are tensor operators (see, e.g., /91/). The effect of Q^0 is absorbed in the first-order spin-orbit parameters. Explicit expressions for the matrix elements of the effective electrostatically correlated spin-orbit interaction for configurations of the type $(d+s)^N$ can be derived by use of common tensor algebra techniques. Results for 1^N-type configurations are given in /90/.

4) The orbit-orbit interaction H_{00}, the spin-spin contact interaction H_{SSC}, and the effect of H_R are absorbed by the first-order and effective electrostatic parameters.

5) The terms which are not absorbed are the spin-dependent spin-spin interaction H_{SS} and the spin-other orbit interaction H_{SOO}, which can be represented by radial integrals P^k, Q^k, and R^k defined by MARVIN /92/ and ARMSTRONG and FENEUILLE /93/. SCHRIJVER and NOORMAN /87/ investigated the influence of the spin-spin and spin-other orbit interactions in some spectra with mixed configurations $(d+s)^N$. They found the influence of these magnetic interactions to be small in comparison to the electrostatically correlated spin-orbit effects.

Using the eigenvectors obtained from the least-squares fit procedure the hfs constants of the real states $|\alpha J>$ can be expressed as linear combinations of the hfs constants between pure SL states $|\tau vSLJ>$,

$$\begin{pmatrix} A(\alpha J) \\ B(\alpha J) \end{pmatrix} = \sum_{\substack{\tau vSL \\ \tau'v'S'L'}} <\tau vSLJ|\alpha J><\alpha J|\tau'v'S'L'J> \begin{pmatrix} A(\tau vSLJ,\tau'v'S'L'J) \\ B(\tau vSLJ,\tau'v'S'L'J) \end{pmatrix} \quad , \tag{45}$$

where $A(\tau vSLJ,\tau'v'S'L'J)$ and $B(\tau vSLJ,\tau'v'S'L'J)$ are proportional to the matrix elements of the hfs operators H_{dip}^{eff} and H_Q^{eff} of (7), respectively. Explicit expressions for these matrix elements for 1^N and $1^{N-1}1'$ configurations are given by CHILDS /21,94/. In the case of a $(d+s)^N$ model space the A and B factors can be expressed as linear combinations of ten $a^{k_s k_l}$ and nine $b^{k_s k_l}$ parameters, respectively, if one neglects the SL-dependent contributions to the $<r^{-3}>^{k_s k_l}$ integrals and the off-diagonal integrals $<d|r^{-3}|s>$. For each of the three configurations three a's and three b's describe the hfs of the d electrons, and for the configuration $d^{N-1}s$ an additional parameter a_s^{10} describes the magnetic dipole interaction of the s electron. By fiting these parameters to the experimental hfs constants experimental values for the $<r^{-3}>$ expectation values are obtained, and according to (31) the difference between these experimental values and theoretical values calculated by relativistic SCF methods can be regarded as a measure of the influence of CI effects on the hyperfine interaction.

2.6 Hyperfine Structure in an External Magnetic Field

The interaction of an atom with an external magnetic field is given by the Zeeman operator

$$H_Z = \mu_B (\underline{L} + g_s\underline{S}) \cdot \underline{H} + \mu_B g_I \underline{I} \cdot \underline{H} \tag{46}$$

where the nuclear g factor is defined by

$$\underline{\mu}_I = -\mu_B g_I \underline{I} \tag{47}$$

and g_s is the free electron g factor. If the field is small enough that J remains a good quantum number, the interaction between the electrons and the external field may be written as

$$\mu_B g_J \underline{J} \underline{H} \tag{48}$$

where the electronic g factor is defined by

$$\underline{\mu}_J = -\mu_B g_J \underline{J} \tag{49}$$

and the total magnetic moment of the electrons is given by

$$\underline{\mu}_J = - \mu_B \langle \alpha J\ M_J=J | \underline{L} + g_s \underline{S} | \alpha J\ M_J=J \rangle \quad . \tag{50}$$

Comparing (49) and (50) we find

$$g_J(\alpha J) = 1 + (g_s - 1) \frac{\langle \alpha J\| \underline{S} \| \alpha \rangle}{\sqrt{J(J+1)(2J+1)}} \quad . \tag{51}$$

For a pure SL state $|\tau v SLJ\rangle$ the reduced matrix element can easily be evaluated. In this case the g_J factor becomes simply

$$g_J(\tau v SLJ) = 1 + (g_s - 1) \frac{J(J+1) - L(L+1) + S(S+1)}{2J(J+1)} \quad . \tag{52}$$

Using the eigenvectors derived from the energy level fitting the g_J factor of a real state $|\alpha J\rangle$ can be expressed as a linear combination of g_J factors of pure SL states,

$$g_J(\alpha J) = \sum_{\tau v SL} |\langle \tau v SLJ | \alpha J \rangle|^2\ g_J(\tau v SLJ) \quad . \tag{53}$$

Then the comparison between experimental and calculated g_J values allows an estimation of the quality of the intermediate coupling wavefunctions. In the heavier atoms relativistic and diamagnetic effects /95-98/ also have to be considered before the calculated g_J values can be compared with precise experimental results. For the rare-earth elements these corrections are of the order of 1 part in 10^3 /99/. In contrast, corrections to the g_J factors caused by the motion of the nucleus /100/ are of the order of 1 part in 10^6 and are negligible compared with the experimental uncertainties.

The $(2F+1)$-fold degeneracy of each hfs level is removed by the Zeeman interaction. The Hamiltonian for the hyperfine interaction in the presence of an external magnetic field can be written as /39/

$$H_{hfs,Z} = hA(J)\underline{I}\cdot\underline{J} + hB(J)\ \frac{3(\underline{I}\cdot\underline{J})^2 + (3/2)(\underline{I}\cdot\underline{J}) - I(I+1)J(J+1)}{2I(2I-1)J(2J-1)} + \mu_B g_J \underline{J}\cdot\underline{H} + \mu_B g_I \underline{I}\cdot\underline{H}. \tag{54}$$

In this expression the magnetic octupole and electric hexadecapole interactions have been neglected; corresponding expressions for these interactions are given in /101/. At low magnetic field, the first two terms dominate, and the Zeeman interaction can be conveniently expressed in the form

$$H_Z = \mu_B g_F \underline{F}\cdot\underline{H} \tag{55}$$

where g_F is given by

$$g_F = g_J \frac{F(F+1) + J(J+1) - I(I+1)}{2F(F+1)} + g_I \frac{F(F+1) + I(I+1) - J(J+1)}{2F(F+1)} \quad . \tag{56}$$

Each hyperfine level splits into $(2F+1)$ equidistant Zeeman components (linear Zeeman effect),

$$E_{FM_F} = <J \, I \, F \, M_F | \, \mu_B \, g_F \, F_z \, H | \, J \, I \, F \, M_F> \, = \, \mu_B \, g_F \, M_F \, H \quad , \tag{57}$$

where we have chosen the z-axis to coincide with the direction of the external field H.

At strong magnetic field, $\underline{I}$ and $\underline{J}$ are decoupled; F is not a good quantum number, but M_J and M_I are. The Zeeman terms of $H_{hfs,Z}$ dominate (Paschen-Back effect), and the energy of the levels is given by

$$E_{M_J M_I} = h \, A(J) \, M_J \, M_I + h \, B(J) \, \frac{\left[3M_I^2 - I(I+1)\right]\left[3M_J^2 - J(J+1)\right]}{2I(2I-1)2J(2J-1)} + \mu_B \, g_J \, M_J \, H + \mu_B \, g_I \, M_I \, H . \tag{58}$$

In a very strong magnetic field also $\underline{L}$ and $\underline{S}$ are decoupled. However, hfs measurements are generally made at fields low enough that we need not consider that effect here.

In general, of course, one is in neither the high-field nor the low-field regime, and a diagonalization of the matrix of the Hamiltonian $H_{hfs,Z}$ must be carried out. Since M_F remains a good quantum number at all fields, the matrix decomposes into a number of submatrices, one for each possible value of M_F. The eigenvalues of the matrices are functions of the field and give the energy of each magnetic substate. The eigenvectors show the composition of each substate in terms of the zero-field basis states of pure F,

$$|\alpha J \, I \mathscr{F} M_F> \, = \, \sum_F <\alpha \, J \, I \, F \, M_F | \alpha \, J \, I \mathscr{F} M_F> \, |\alpha \, J \, I \, F \, M_F> \quad . \tag{59}$$

The quantity $\mathscr{F}$ is written in script to indicate that it is not a good quantum number at a non-zero field.

The Zeeman interaction provides the possibility of determining the nuclear magnetic dipole moments independently of the interaction between the nucleus and the atomic electrons by extracting, from adequate hfs measurements, the direct interaction between the nuclear magnetic dipole moment and the external magnetic field. However, in order to obtain the correct value of the magnetic moment from such measurements, one must first correct for the distortion of the atomic core by the external magnetic field /102/. The external field induces a diamagnetic current density in the atomic core which produces a magnetic field H'(0) at the nucleus. This induced field is opposed to the external field. Thus, if an external field H is applied, the field seen by the nucleus is not H, but rather

$$H \left[1 + \frac{H'(0)}{H} \right] = H(1-\sigma) \quad . \tag{60}$$

This implies that the magnetic moments determined from their interaction with the external magnetic field must be multiplied by a factor $1/(1-\sigma)$ to correct for diamagnetic shielding. Values for the correction factor based on the relativistic Hartree-Fock-Slater calculations of FEIOCK and JOHNSON /103/ are tabulated in /1/.

2.7 Off-Diagonal Hyperfine and Zeeman Interactions

In the previous sections we have assumed that J is a good quantum number, in particular, the form of the Hamiltonian $H_{hfs,Z}$ of (54) depends on this assumption. However, there can exist hyperfine as well as Zeeman interactions between the state under investigation $|\alpha J\rangle$ and other atomic states $|\alpha'J'\rangle$. If the hyperfine energies are assumed to be small compared to the fine-structure splitting, the energy shift of a hyperfine level $|\alpha J\,I\mathscr{F}M_F\rangle$ due to these off-diagonal interactions can be calculated using second-order perturbation theory /21/,

$$\delta E(\alpha J I \mathscr{F} M_F) \cong \sum_{(\alpha J) \neq (\alpha'J')} \sum_{F'} \frac{|\langle \alpha J I \mathscr{F} M_F | H_{dip} + H_Q + \mu_B(\underline{L} + g_s \underline{S}) \cdot \underline{H} | \alpha'J'IF'M_F\rangle|^2}{E(\alpha J) - E(\alpha'J')} \tag{61}$$

where magnetic octupole and electric hexadecapole interactions have been neglected. Since only matrix elements diagonal in I are considered, the nuclear term of the Zeeman interaction can be neglected, too. Explicit expressions for the off-diagonal matrix elements in (61) are given in /21,94/. For the 4d- and 5d-shell atoms the perturbations have been evaluated within the three-configuration model space $(d+s)^N$, but off-diagonal matrix elements between the configurations have been neglected.

Evaluating the parameters which appear in the effective hyperfine and Zeeman Hamiltonian, namely $a^{k_s k_l}$, $b^{k_s k_l}$, g_J, and g_I, by fitting experimental data can in principle be done by assuming that J is a good quantum number and using a Hamiltonian which contains effective operators to correct for second-order mixing of atomic states /39/. In practice, however, one uses an iterative technique. One calculates the unknown parameters from the experimental data ignoring second-order interactions, then one uses these approximate parameter values·to evaluate the second-

order corrections according to (61). The second-order corrections are then subtracted from the measured values. A new fit leads to improved parameter values which can
be used to calculate better second-order corrections, etc.

From (61) it can be seen that we may distinguish three second-order correction
terms, one containing products of two hfs matrix elements, a second one containing
products of two Zeeman matrix elements, and a mixed one containing products of a
hfs and a Zeeman matrix element. The influence of the "hfs-hfs" term can be taken
into account by using effective values of the interaction constants instead of the
corrected values derived from the data corrected for off-diagonal effects /104/. The
effective hfs constants are determined from measurements of the hyperfine splitting
at magnetic fields which are low enough that the field-dependent correction terms can
be neglected. On the other hand, the "field-field" and "hfs-field" corrections play
an important role in the determination of the g factors g_J and g_I from high-field
measurements (Sect.7).

In principal, all levels of the model space which may contribute to δE according
to the selection rule $|\Delta J| \leq 2$ should be considered as perturbing states. This leads
in the case of the $(d+s)^N$ model spaces of the 4d- and 5d-shell atoms to calculations
which are expensive in computer time due to the strong spin-orbit and configuration
mixing in these atoms. Therefore, the number of perturbing levels has to be limited.
In many cases, the members of the same SL multiplet as the perturbed state and other
nearby states give the largest contributions to the second-order corrections. However, even levels which lie more than 10000 cm^{-1} away from the perturbed state may
contribute appreciably to the corrections. Thus, the question which levels should
be included in the second-order calculation has to be examined carefully for each
individual case.

2.8 Hyperfine Anomaly

The magnetic dipole interaction constant was defined by (3,5,8),

$$A = \frac{\mu_I}{h I J} <J J |T_0^1(e)|J J > \quad . \tag{62}$$

According to the Hamiltonian (10), the electronic wavefunctions depend on only one
nuclear quantity, the total charge Z. Thus, the electronic matrix element
$<J J |T_0^1(e)|J J>$ should be the same for two isotopes of the same element. This leads
to the relationship

$$\frac{A(1)}{A(2)} = \frac{\mu_I(1)/I(1)}{\mu_I(2)/I(2)} = \frac{g_I(1)}{g_I(2)} \quad . \tag{63}$$

This expression can be used to determine the nuclear magnetic dipole moment from the magnetic dipole interaction constant A if the magnetic moment and the A factor have been measured independently for a reference isotope. Deviations from this relationship are expressed in terms of the hyperfine anomaly ${}^1\Delta^2$,

$$^1\Delta^2 = \frac{A(1) \, g_I(2)}{A(2) \, g_I(1)} - 1 \quad . \tag{64}$$

The anomaly, although generally smaller than 1%, can be almost as large as 10% /105,106/.

There are two effects contributing to the magnetic dipole hyperfine anomaly which are both caused by the finite extension of the nucleus. The first contribution, the Breit-Rosenthal effect /107/, is due to the fact that the nuclear charge has extension. Inside the nuclear volume the potential deviates very much from the Coulomb potential Ze/r. Therefore the electronic radial wavefunctions obtained using a potential appropriate to an extended nucleus differ from the wavefunctions obtained for a point nucleus. Consequently, there is a correction ε_{BR} to the point hfs that results from the extended nuclear charge. The Breit-Rosenthal effect can reach a considerable amount for large Z, but its isotopic variation due to isotopic changes in the nuclear volume is very small and can be estimated using various models for the nuclear charge distribution /108/. The second contribution to the hyperfine anomaly takes into account the extended nuclear magnetization. The factorization of the hfs (3) was based on the assumption that the electrons are always outside the nucleus, that means, that the nuclear magnetization has been approximated by $\mu_I \delta^3(0)$. Clearly, the A factor of (62) has to be modified. The fractional difference between the point nucleus magnetic dipole interaction and the one that is obtained for the extended nuclear magnetization is known as the Bohr-Weisskopf anomaly ε_{BW} /109/. Moreover, in the calculation of ε_{BW} electronic wavefunctions have to be used which correspond to the actual nuclear charge distribution. Thus the extended nuclear charge also affects the Bohr-Weisskopf anomaly. The isotopic variations of the magnetic moments, combined with the different contributions to the hfs of the orbital and spin parts of the magnetization in the case of the extended nucleus, allow for relatively large isotopic variations in the deviation from the point dipole interaction. Therefore, the Bohr-Weisskopf effect is very often the dominating part of the hyperfine anomaly. Since especially $s_{1/2}$ electrons (and to less extent $p_{1/2}$ electrons) penetrate the nucleus, contributions to the hyperfine anomaly may arise from open s-electron (or p-electron) shells or from s-electron admixtures to the electronic wavefunction due to spin polarization. The s-electron contact part A_s of the A factor can then be written as

$$A_s = A_{point} \, (1 + \varepsilon_{BW})(1 + \varepsilon_{BR}) \tag{65}$$

where A_{point} is the A_s factor in the point nucleus approximation and contains not only the contribution from the contact interaction parameter a_s^{10} of the unpaired s electron in the configuration $d^{N-1}s$, but also those contributions from the parameters a_d^{10} which arise from spin polarization.

Using the experimental A factor and the contact part A_s, the s-electron hyperfine anomaly $^1\Delta_s^2$ may be extracted from the experimental hyperfine anomaly $^1\Delta_{exp}^2$ /7/,

$$^1\Delta_s^2 = \frac{A_{exp}}{A_s} \, ^1\Delta_{exp}^2 \quad . \tag{66}$$

Using (65) we find

$$^1\Delta_s^2 = \frac{\left[1 + \varepsilon_{BW}^{(1)}\right]\left[1 + \varepsilon_{BR}^{(1)}\right]}{\left[1 + \varepsilon_{BW}^{(2)}\right]\left[1 + \varepsilon_{BR}^{(2)}\right]} - 1 \quad , \tag{67}$$

and for the isotope shift of the Bohr-Weisskopf correction

$$^1\Delta_{BW}^2 = \varepsilon_{BW}^{(1)} - \varepsilon_{BW}^{(2)} = \, ^1\Delta_s^2 \, (1 + \varepsilon_{BW}^{(2)}) - \frac{1 + \varepsilon_{BW}^{(1)}}{1 + \varepsilon_{BR}^{(2)}} \, ^1\Delta_{BR}^2 \quad . \tag{68}$$

The comparison of experimentally determined Bohr-Weisskopf anomalies with theoretical predictions provides detailed information on nuclear structure. Several nuclear models have been used to evaluate the Bohr-Weisskopf correction resulting in satisfactory agreement with experimental results /109-115/. MOSKOWITZ and LOMBARDI /116/ have established an empirical rule for the Bohr-Weisskopf correction which fits the hyperfine anomalies for ten mercury isotopes very well,

$$\varepsilon_{BW} = \frac{\alpha}{\mu_I} \tag{69}$$

where for s-electron contributions $\alpha = \pm 1.13 \cdot 10^{-2} \, \mu_N$ for $I = 1 \pm 1/2$, 1 being the orbital angular momentum of the odd neutron. For odd protons opposite signs have to be used. FUJITA and ARIMA /117/ gave a theoretical interpretation of this rule by taking effects of nuclear core polarization and mesonic exchange currents into account. The application of this rule to the case of odd-odd nuclei has been discussed by EKSTRÖM and co-workers /118/

Since application to iridium /119/ and gold /120/ isotopes yields hyperfine anomalies which also agree very well with the experimental ones, it appears that for this region of almost spherical nuclei around $Z = 80$ the Moskowitz-Lombardi rule may be used to estimate the Bohr-Weisskopf correction.

3. Experimental Methods

3.1 Atomic Beam Magnetic Resonance

The transition elements with an unfilled 4d or 5d electron shell have many low-lying metastable atomic states belonging to the three configurations $(d+s)^N$, which are, in a thermal atomic beam, sufficiently populated for hfs measurements with the ABMR method. In addition, several experimental techniques have been developed which also allow the investigation of very weakly populated high-lying metastable states by ABMR. In order to enhance the population of metastable states electron bombardment of the atomic beam /121/, plasma discharge metastabilization /122/, and optical pumping using a cw dye laser /123/ have been used. Efficient detection of high-lying metastable atoms based on Auger de-excitation /121/ has been applied in many experiments. Recently, a detection technique using a tunable single-mode cw dye laser has brought considerable progress because of its state-selectivity and high efficiency (Sect.3.3) /66,124/.

The greater part of the hfs investigations of 4d- and 5d-shell atoms has been performed by the ABMR group at Bonn using a 100-kV electron-bombardment atomic beam source that allows the production of intense atomic beams of these highly refractory elements. The ABMR method, devised by RABI and co-workers /15,16/ in 1938-1940, has been described in detail many times /17-22/. Therefore, only the details relevant to the experiments on the 4d- and 5d-shell atoms will be described here.

Figure 2 shows schematically the ABMR apparatus used. The strong inhomogeneous magnetic fields A and B, in which the atoms evaporated from the atomic beam source are deflected due to their effective magnetic moment $\mu_{eff} \cong \mu_B g_J M_J$, are of the two-wire type. The gradient of the B magnet is in the same sense as that of the A magnet. Thus, transitions between the Zeeman levels of the hfs, which are induced in the homogeneous C-field region by an rf magnetic field, can be detected as "flop-in" signals if $M_J(B) = -M_J(A)$.

Transitions with $M_J(B) \neq M_J(A)$ which do not fulfil this symmetry condition can be detected as "flop-in" signals, too. For this purpose the atomic beam source has to be adjusted unsymmetric off the ABMR machine axis /125/. This technique is particu-

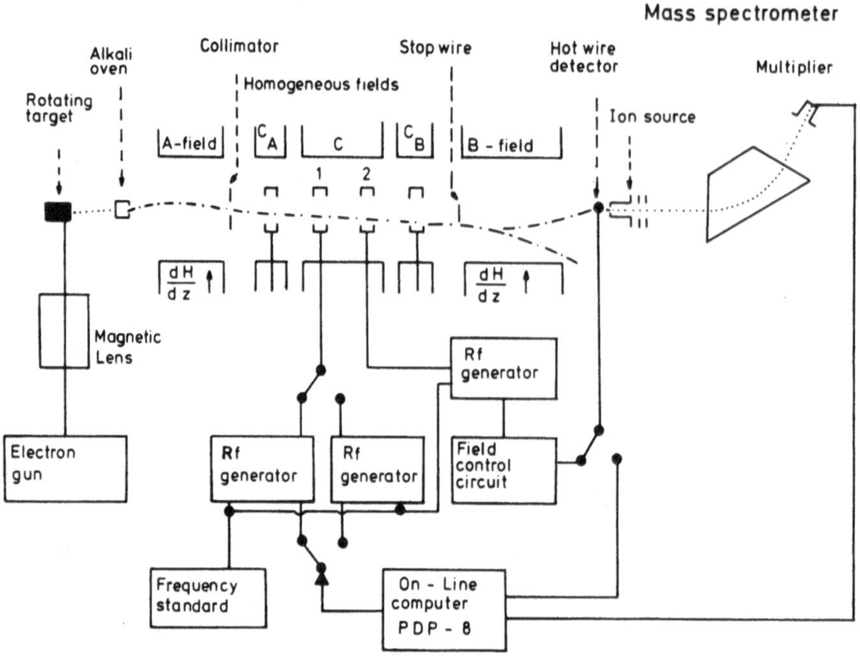

Fig.2. Schematic diagram of the ABMR apparatus at Bonn equipped with a universal detector

larly useful for atomic states with integral J value. Symmetric adjustment of the apparatus allows detecting transitions of the type $M_J(A)=0 \rightarrow M_J(B)\neq0$ as "flop-out" signals; since the atoms in a state with $M_J=0$ pass through the apparatus without any deflection, such a transition results in a small decrease of the intensity observed at the detector. However, detecting such transitions as "flop-in" signals leads to a much better signal-to-noise ratio because the background of non-resonant atoms is much smaller in this case.

Transitions of the type $\Delta M_J=0$ are not detectable directly, either as "flop-out" or as "flop-in" signals, but can be detected by applying the triple resonance technique, i.e., by inducing two supplementary rf transitions in two homogeneous magnetic fields C_A and C_B which are separated from the original C-field (Sect.7.1).

In order to obtain a suitable long-term stability the homogeneous C-field is controlled by observing a strongly field-dependent transition in a simultaneously running alkali beam, which is detected by a hot-wire detector. For this purpose the alkali reference transition signals are fed into a control system which keeps the C-field strength at a constant value H_0 by locking the resonance to a fixed frequency /126/.

For detecting the investigated 4d- and 5d-shell atoms a so-called universal detector has been used. The atoms which pass through the detector slit are ionized by electron bombardment, mass separated by a 60° mass spectrometer, and counted by a multiplier tube. In addition, during the investigation of high-lying metastable states of rhenium advantage has been taken of a state-selective detection method using a single-mode cw dye laser (Sect.3.3).

The rf resonances were recorded using a PDP-8 on-line computer system. Two experimental procedures have been used controlled by the programs C08 /127/ and M04 /128/, respectively.

In the C08 procedure the resonance curves are scanned in small equidistant frequency steps. The frequency is stepped through the relevant frequency range forward and backward many times, and the counting rate differences $(N_+ - N_-)$, where N_+ and N_- are the counting rates for the rf power switched on and off, respectively, are summed up. Then, the resonance frequency is determined by fitting the resonance curve to a Gauss profile. Using this procedure the resonance frequencies can be measured with an accuracy of about 10% of the linewidth, which is typically 50-80 kHz (FWHM).

Because the rf transition to be measured and the alkali reference transition used for the C-field control are induced in neighboring rf loops, the magnetic field strengths for the two transitions differ slightly due to inhomogeneities of the magnetic field. The local field strength for the transition to be measured, $H = H_0 + dH$, can be determined by interchanging the positions where the two transitions are induced. If the calibration frequency is held at a constant value the difference of the field strength dH between the two rf loop positions may be determined from the change $d\nu$ in the resonance frequency of the transition to be measured under the assumption that the field profile is unchanged,

$$dH = \frac{d\nu}{2\partial\nu/\partial H} \quad .$$
(70)

In many cases the local field strength was determined by inducing a strongly field-dependent transition in an atomic state of the isotope under investigation, for which the hfs constants and the g_J factor had already been measured, in the same rf loop as the transition to be measured.

In the M04 procedure the transition frequencies are determined by scanning the resonance curves at three points. By dividing up the total counting time into many short counting intervals the influence of fluctuations of the atomic beam intensity can be eliminated to a high degree (signal averaging).

This procedure allows the determination of the resonance frequencies with an accuracy of better than 1% of the linewidth. Two resonance frequencies ν_1 and ν_2 of the same isotope, but not necessarily in the same fine-structure state, are measured alternately several times in the same rf loop, that means, at the same magnetic field H. Then, the local field-independent quantity

$$dv = (v_1 - v_1(H_0)) - \frac{\alpha_1}{\alpha_2} (v_2 - v_2(H_0)) \tag{71}$$

can be constructed, where $v_i(H_0)$ is the transition frequency calculated for the magnetic field H_0 that corresponds to the alkali calibration frequency, and $\alpha_i = \partial v_i / \partial H$ is the field dependence of the transition frequency v_i. dv depends on the hfs constants and on the g factors g_J and g_I. Thus, these quantities can be determined in a least-squares fit procedure if a sufficient number of suitable frequency pairs (v_1, v_2, H) have been measured /101/.

Because of the high precision achieved with the MO4 procedure, systematic shifts to the resonance frequencies have to be eliminated. In order to avoid errors due to the Millman effect /129/, caused by a non-constant direction of the rf field within the transition region, and to phase shifts within the rf loop /130/, each measurement is repeated with the magnetic field reversed and the rf loop inverted, and the transition frequencies are taken as mean values of a complete set of measurements with the four possible combinations of rf loop orientation and field direction. In addition, the resonance frequencies are corrected for shifts due to non-resonant transitions[3] /126/ and to the deviation of the local frequency standard from the atomic time system AT1. For the latter purpose the local standard frequency system is continuously compared with the standard frequency transmitters DCF-77 (Mainflingen, FRG) and GBR-Rugby (Great Britain) /131/.

In order to measure the electronic g_J factors for elements which have a stable isotope with nuclear spin $I = 0$, the Zeeman transitions $M_J = 0 \rightarrow M_J = \pm 1$ or the double quantum transition $M_J = +1 \leftrightarrow M_J = -1$ were measured at high magnetic fields. The g_J factors are determined from the relation

$$v = \frac{\mu_B}{h} g_J H \quad . \tag{72}$$

In the case of elements for which no stable isotope with $I = 0$ exists, the g_J factors are determined from $\Delta F = 0$, $\Delta M = \pm 1$ transitions. At high magnetic fields the increasing dependence of these transitions upon the hfs intervals also allows rough predictions of the hfs constants.

In order to determine the hfs constants with high precision, $\Delta F = 1$, $\Delta M = 0, \pm 1$ transitions are measured at low magnetic fields (typically 0.5-20 Oe). However, the magnetic field must be so large that neighboring resonances are separated sev-

---

[3] The program for the calculation of these corrections was kindly supplied by
W. Foerster, Bonn.

eral line-widths from the resonance to be measured. The influence of the uncertainty in the nuclear g factor g_I can usually be neglected at low magnetic fields.

For the precise determination of the g_I factor from the direct interaction between the nuclear magnetic dipole moment and the external magnetic field, transitions of the type $\Delta M_I = \pm 1$, $\Delta M_J = 0$, which require application of the triple resonance technique, generally have to be measured at high magnetic fields (Sect.7.1).

3.2 Production of Atomic Beams

Atomic beams suitable for study with the ABMR method are usually produced by evapo-
rating the element to be studied from a crucible through a narrow rectangular slit
(for apparatus with two-pole deflecting magnets) or through a small circular hole
(for apparatus with six-pole deflecting magnets) /132/. The heating of the crucible
is usually achieved by resistance heating or, for higher evaporation temperatures,
by electron impact. A fairly common difficulty in the production of suitable beams
of certain elements is that of finding a crucible material that does not react with
the element to be evaporated. In most cases such reactions destroy the crucible and
prevent the production of an atomic beam. These difficulties become almost insur-
mountable for highly refractory elements with evaporation temperatures[4] above 2800 K.
There is only one known example of the production of an atomic beam with this method
above this temperature, namely the evaporation of carbon from a crucible of tantalum
carbide /133,134/. The bulk of the 4d- and 5d-shell elements belong to this class of
highly refractory elements as can be seen from the vapor pressure data in Fig.3.

Consequently, several other methods for the production of atomic beams of refractory elements have been developed. DOYLE and MARRUS /136/ investigated radioactive
Ta, W, Re, and Ir isotopes which were produced by neutron irradiation of thin wires
of these elements. The wires were heated by electron bombardment to just below the
melting point. The resultant intensity of the beams of sublimated atoms was sufficient for the determination of the nuclear spins with the ABMR method because of the
high detection efficiency for radioactive isotopes. Moreover, using this wire-bombard-
ment beam production technique, it was possible to measure the ground-state hfs /137/
and the nuclear magnetic dipole moments /138/ of the two isotopes [186]Re and [188]Re.

A similar method for the production of atomic beams of refractory elements has
been developed by PENDLEBURY and co-workers /139,140/ and was used successfully in

---

[4] Here, by evaporation temperature we mean the temperature at a vapor pressure of
0.1-mm Hg. This is the typical vapor pressure at which ABMR experiments are run.

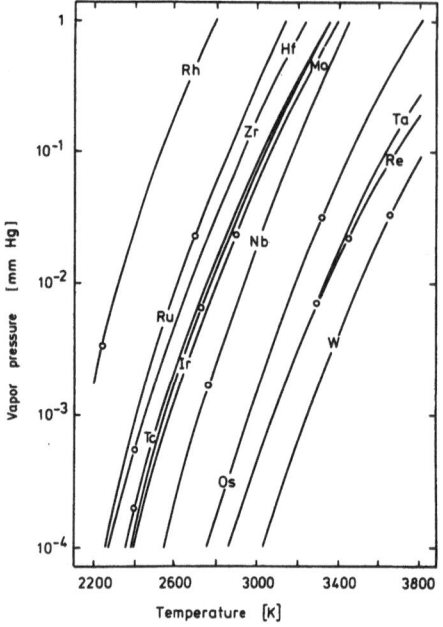

Fig.3. Vapor pressure data for the refractory 4d- and 5d-shell elements /135/. The symbol o indicates the melting point

the investigation of the stable Mo and Re isotopes. They fed a thin wire of the material to be evaporated down through a vertical guiding cylinder. At the end of the wire they produced a drop of molten material by heating this end above the melting point by electron impact. The drop is suspended at the end of the solid wire and atoms evaporate from it. The beam cross section is defined by a narrow slit positioned in front of the drop. The beam intensity was typically about 1/50 of that obtained by conventional evaporation from a crucible, but was sufficient to measure the ground-state hfs of ^{95,97}Mo and ^{185,187}Re. This molten-ball beam-producing technique has also been used by LINDGREN and co-workers /141-143/ in the investigation of a number of radioactive isotopes by ABMR and by MEISEL and co-workers /144/ in the investigation of metastable states of stable Mo isotopes in a laser-rf double-resonance experiment.

Finally, a universal method for the production of atomic beams of stable refractory elements was developed by the ABMR group at Bonn /33/. The principle of the method is shown in Fig.4. An area equivalent in its dimensions to the slit of a conventional atomic beam crucible is heated locally on the mantle surface of a cylindrical target consisting of a solid piece of the element to be studied. The local heating above the evaporation temperature is achieved by the well-collimated electron beam of a 100-kV electron gun. The electron beam can be focussed by a magnetic lens to a few tenths of a millimeter. In order to obtain an evaporation zone of rectangular shape with dimensions of about 0.3×5 mm, which is the required cross section of the atomic beam, the electron beam is deflected with a frequency

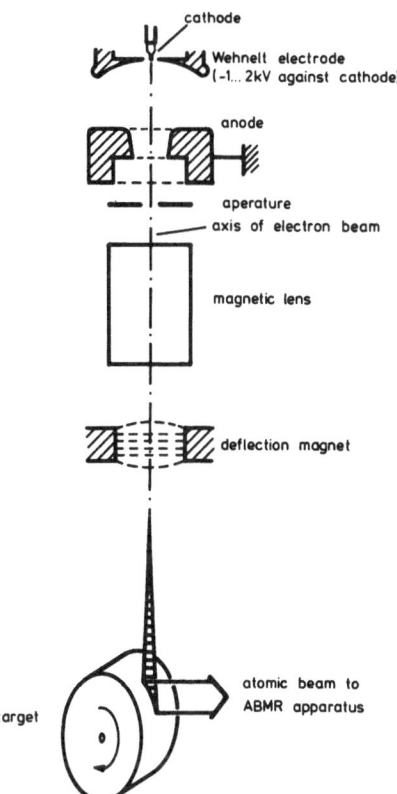

◄**Fig.4.** Schematic diagram of the rotating-target atomic beam source

Fig.5. Cross section of the rotating-target mounting

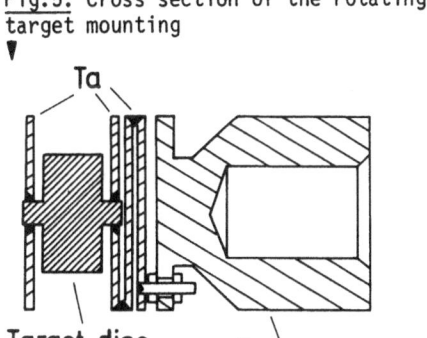

of 25 kHz linearly in one direction by a deflection magnet. The geometrical shape of the atomic beam passing through the ABMR apparatus is thus defined by the evaporation zone, the collimating slit in the middle of the apparatus, and the entrance slit of the detector (Fig.2). The cylindrical target with a diameter of 30 mm and a thickness of 15 mm is mounted, together with a suitable heat-shielding arrangement made of tantalum, onto a water-cooled brass block (Fig.5). In order to avoid deterioration of the target surface, the target is rotated at a slow angular speed of about 1 revolution in 50 s, allowing the material to cool down below the melting point soon after it has left the area hit by the electron beam. By very slowly moving the target back and forth in the direction of the axis the whole mantle surface of the target is gradually used for the evaporation process.

Using this rotating-target beam production technique atomic beams of all highly refractory elements can be produced. The resultant beam intensities are comparable to those obtained by evaporation from a crucible. However, atomic beam intensity fluctuations of up to 20% are typical for this technique, thus requiring signal-averaging methods for the detection of rf resonances (Sect.3.1). This technique has been used by the ABMR group at Bonn to perform hfs and Zeeman studies on meta-

40

stable states of the following elements: Zr, Nb, Mo, Ru, Hf, Ta, W, Re, Ir, and Pt.
For the production of atomic beams of these elements electron currents between 3 and
12 mA were necessary, corresponding to a power density of between 200 and 800 W/mm^2
on the target.

3.3 High-Resolution Laser Spectroscopy and State-Selective Detection of Metastable
Atoms

In the course of investigating 4d- and 5d-shell atoms metastable states with excitation energies up to 17000 cm^{-1} were found in some cases to be sufficiently populated to detect rf resonances. Although the population of the states is possibly
enhanced by electron impact achieved by the evaporation technique (see also /145/),
the signal-to-noise ratio of the resonance curves decreases with increasing excitation energy. In addition, the background of a conventional "flop-in" signal of
an ABMR rf resonance, which consists of atoms of all metastable states and of the
ground state, also makes the signal-to-noise ratio worse. Thus, up to 20 min of
data collection were necessary to obtain useful resonance curves in the higher lying states (Fig.6a).

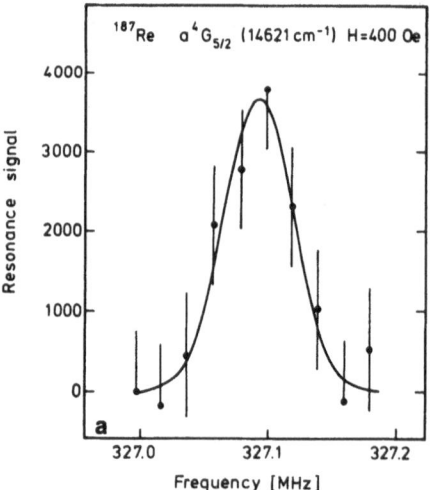 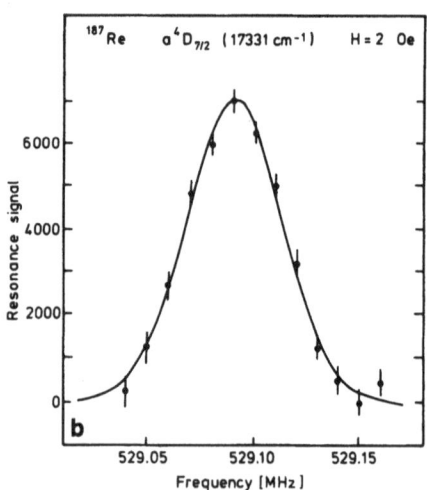

Fig.6a. Hfs transition in the $^4G_{5/2}$ state of ^{187}Re detected with the universal detector. The resonance curve with a signal-to-noise ratio of about 5:1 required 1000 s
of data collection. b) Hfs transition in the $^4D_{7/2}$ state of ^{187}Re detected with the
laser detector. The resonance curve with a signal-to-noise ratio of about 25:1 required only 70 s of data collection

With the advent of tunable dye lasers it has become possible to achieve selective excitation of metastable states /123/ as well as selective detection of metastable atoms /124,146,147/. A modification of the conventional ABMR method, the laser-rf double-resonance method, that replaces the inhomogeneous deflecting magnets by two laser beams, has been developed at Bonn by ERTMER and HOFER /145/ and independently at London (Canada) by ROSNER et al. /148/.

Recently, a state-selective detection system using a tunable single-mode cw dye laser has been developed for the Bonn ABMR apparatus and was used successfully in the investigation of weakly populated high-lying metastable states of Re /66/. Figure 7 shows schematically the ABMR apparatus with the laser detection system. The dye laser system consists of a Spectra Physics model 580A single-mode dye laser pumped by an Ar^+ laser. Part of the light is split off and fed to the frequency control system. The dye laser is locked to a confocal interferometer, which is kept constant by referencing it to one of the two rf-generated side-bands of an iodine stabilized He-Ne laser. The dye laser is precisely tuned by changing the frequency of the rf source /149,150/.

The light from the dye laser is irradiated at right angle to the atoms passing through the detector slit of the ABMR apparatus. Fluorescence light released after the excitation of the metastable atoms to an upper level of opposite parity is detected in a direction perpendicular to the atomic beam as well as to the laser beam.

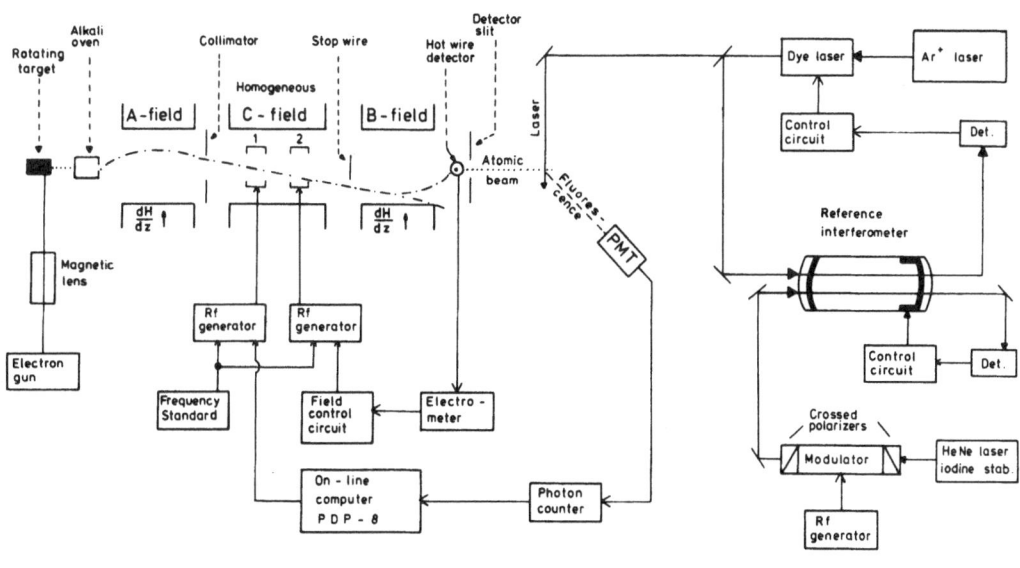

ABMR apparatus   Dye laser system

Fig.7. Experimental set-up of ABMR with laser detection

Since the frequency jitter of the laser system is well below 1 MHz, hfs and IS of
the spectral line are resolved and the photomultiplier signal is proportional to
the number of atoms in an individual hyperfine level of an individual isotope which
pass through the detector slit, i.e., the number of resonant atoms. Because of the
high efficiency of this detection method much shorter periods of data collection
are necessary to obtain a useful signal-to-noise ratio. (Fig.6b).

In a first step of the Re experiment the hfs of 14 metastable states was measured
by high-resolution fluorescence spectroscopy. From these measurements the hfs constants A and B of the lower metastable states and of the upper states as well as
the IS of the optical lines were determined with an accuracy of typically 1 part

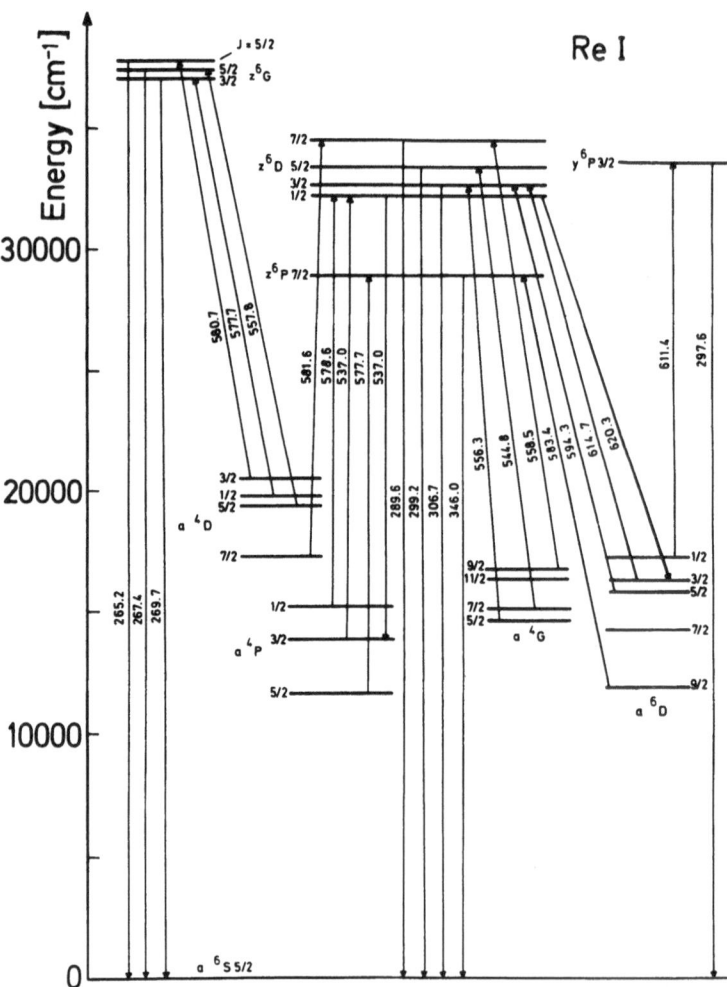

Fig.8. Energy level diagram for ReI

in 10^{-3} for the A factors, 2 parts in 10^{-2} for the B factors, and 5 parts in 10^{-3} for the IS.

In a second step of the experiment the hfs of the levels $5d^5 6s^2$ $^4D_{7/2}$ and $5d^6 6s$ $^6D_{1/2,3/2,5/2,9/2}$ was measured by ABMR. For this purpose the dye laser was locked to an individual hfs component of the transition to the upper level. Since rough values for the hfs intervals were known from the fluorescence measurements, the rf transitions were easily found. In Fig.8 an atomic energy level diagram of the metastable states up to an excitation energy of 21000 cm^{-1} is given together with the laser-induced transitions which connect the investigated metastable levels with the upper states of opposite parity. The spectral lines used for the fluorescence light detection are also indicated in the figure.

4. Experimental Hyperfine Interaction Constants for 4d- and 5d-Shell Atoms

Tables 1 and 2 summarize the hfs constants A and B for stable 4d- and 5d-shell iso-topes, which have been measured by ABMR. In order to be able to determine the effec-tive hfs radial parameters for as many elements as possible, in some cases hfs data obtained by classical optical spectroscopy (O), laser induced fluorescence (LIF), optogalvanic spectroscopy, and laser-rf double-resonance (L/RF) have been added to the ABMR results.

Corrections for off-diagonal hyperfine interactions[5] (Sect.2.7) are given in columns 5 and 7 of the tables if available. In order to obtain the corrected A and B constants, the corrections δA and δB have to be subtracted from the experimental values A_{exp} and B_{exp}. It can be seen from Tables 1 and 2 that the off-diagonal cor-rections are very often much greater than the experimental errors. Usually, the change in B is greater than that in A. Especially for states lying close in ener-gy, the correction δB reaches relatively large values up to several MHz.

---

[5] The corrections have been evaluated using the computer program PERJ. The basic version of this program, which had to be extended for the present calculations to the case of a $(d+s)^N$ model space, has been written by L.S. Goodman and W.J. Childs, Argonne National Laboratory. For elements investigated by other authors the corrections calculated by these authors are given.

Table 1. A and B factors for 4d-shell atoms. The level designations and energies are taken from /151/

Isotope	Configuration and state	Level energy [cm⁻¹]	A_{exp} [MHz]	δA [kHz]	B_{exp} [MHz]	δB [kHz]	Method	Reference
^{89}Y	$4d5s^2$ $^2D_{3/2}$	0	-57.217 (15)				ABMR	152
	$^2D_{5/2}$	530	-28.749 (30)					
^{91}Zr	$4d^25s^2$ $^3F_2$	0	-170.7696 (38)	0.0	-21.502 (29)	-2	ABMR	153
	$^3F_3$	570	-104.9442 (38)	-0.1	-24.184 (49)	3		
	$^3F_4$	1241	-77.8878 (38)	0.0	-33.477 (62)	-1		
	$4d^35s$ $^5F_2$	5023	-105.7511 (48)	-24.7	5.271 (36)	-316		
	$^5F_3$	5249	-189.1516 (89)	-10.3	9.400 (113)	-206		
	$^5F_4$	5541	-217.5756 (80)	-3.6	17.217 (131)	-103		
	$^5F_5$	5889	-229.6592 (58)	-0.2	27.672 (100)	-12		
^{93}Nb	$4d^35s^2$ $^4F_{3/2}$	1143	644.16 (54)ᵃ		32.8 (12.8)ᵃ		ABMR	154
	$^4F_{5/2}$	1587	372.4692 (17)	-16.1	32.658 (38)	-683		
	$^4F_{7/2}$	2154	292.1939 (16)	-8.3	44.320 (47)	-608		
	$^4F_{9/2}$	2805	269.6299 (26)	-2.1	63.805 (130)	-510		
	$4d^45s$ $^6D_{1/2}$	0	1868.1052 (18)	-113.2				
	$^6D_{3/2}$	154	852.4581 (21)	-84.7	-68.948 (41)	-4376		
	$^6D_{5/2}$	392	719.4488 (18)	-27.1	-49.596 (39)	-1991		
	$^6D_{7/2}$	695	690.1932 (16)	-4.6	20.195 (75)	-280		
	$^6D_{9/2}$	1050	691.6192 (22)	5.1	134.064 (116)	1310		
^{95}Mo	$4d^45s^2$ $^5D_1$	11143	-40.817 (3)	-5	-2.995 (8)	-35	L/RF	144
	$^5D_2$	11454	-50.6417 (10)	-1.8	-2.131 (7)	-18	LIF	155
	$^5D_3$	11859	-64.7 (7)		8 (8)			

Table 1. (cont.)

Isotope	Configuration and state	Level energy $[cm^{-1}]$	A_{exp} [MHz]	δA [kHz]	B_{exp} [MHz]	δB [kHz]	Method	Reference
97Mo	$4d^55s$ $^5D_4$	12346	-75.3 (8)		12 (12)			
	$^7S_3$	0	-208.58206 (1)	0.50	0.03705 (10)	20.13	ABMR	156
	$^5S_2$	10768	429.4 (1.0)				Optog.	157
	$4d^45s^2$ $^5D_1$	11143	-41.676 (3)	-5	33.983 (6)	-33	L/RF	144
	$^5D_2$	11454	-51.7015 (16)	-2.4	24.306 (11)	-23	LIF	155
	$^5D_3$	11859	-66.8 (9)		-10 (9)			
	$^5D_4$	12346	-77.5 (1.2)		-83 (15)			
99Tc	$4d^55s$ $^7S_3$	0	-212.98093 (1)	0.52	-0.06999 (14)	20.79	ABMR	156
	$4d^55s^2$ $^6S_{5/2}$	0	-109.1 (9)				O	158
	$4d^65s$ $^6D_{1/2}$	4179	1888.7 (15.0)					
	$^6D_{3/2}$	4003	894.9 (3.0)		63 (15)			
	$^6D_{5/2}$	3701	772.0 (4.5)		0 (30)			
	$^6D_{7/2}$	3251	761.5 (4.5)		-48 (30)			
	$^6D_{9/2}$	2573	813.9 (4.5)					
	$^4D_{1/2}$	11891	2395.3 (6.0)					
	$^4D_{3/2}$	11579	51.0 (6.0)					
	$^4D_{5/2}$	11063	-215.9 (6.0)					
	$^4D_{7/2}$	10517	-389.7 (6.0)					
99Ru	$4d^75s$ $^5F_2$	2713	-82.5290 (27)	3.2	5.489 (22)	25		159,160
	$^5F_3$	2092	-135.0279 (37)	1.7	10.180 (50)	13	ABMR	
	$^5F_4$	1191	-163.68791 (12)	1.06	17.4760 (15)	4.2		160

Isotope	Config.	State		A (MHz)			B (MHz)		Method	Ref.
^{101}Ru	$4d^7 5s$	$^5F_5$	0	−204.55595 (8)	0.63	27.3206 (9)		42.5	ABMR	159,160
		$^5F_2$	2713	−92.4929 (27)	4.1	31.916 (23)		37		
		$^5F_3$	2092	−151.3482 (38)	2.1	59.349 (50)		21		160
		$^5F_4$	1191	−183.47766 (12)	1.34	101.8178 (11)		8.3		
		$^5F_5$	0	−229.29149 (11)	0.75	158.9813 (13)		54.7		
^{103}Rh	$4d^8 5s$	$^4F_{3/2}$	3473	32.377 (1)					ABMR	161
		$^4F_{5/2}$	2598	−107.385 (2)						
		$^4F_{7/2}$	1530	−87.416 (1)						
		$^4F_{9/2}$	0	−175.574 (1)						
		$^2F_{5/2}$	7791	−210.881 (323)						
		$^2F_{7/2}$	5691	−88.198 (1)						
	$4d^9$	$^2D_{3/2}$	5658	−91.718 (1)						
		$^2D_{5/2}$	3310	−87.364 (2)						
^{105}Pd	$4d^9 5s$	$^3D_1$	10094	76.6 (6)[b]					ABMR	162
		$^3D_2$	7755	66.359 (1)		−398.192 (10)				
		$^3D_3$	6564	−391.178 (1)		−652.906 (15)				
		$^1D_2$	11722	−621 (5)		−490 (30)				
^{107}Ag	$4d^9 5s^2$	$^2D_{5/2}$	30242	−126.2818 (1)	0.0				ABMR	121
	$4d^{10} 5s$	$^2S_{1/2}$	0	−1712.512111 (18)						163
^{109}Ag	$4d^9 5s^2$	$^2D_{3/2}$	34714	−365.2 (6.9)					0	164
		$^2D_{5/2}$	30242	−145.1584 (5)	0.0				ABMR	121
	$4d^{10} 5s$	$^2S_{1/2}$	0	−1976.932075 (17)						163

[a] Corrected A and B factors determined from high-field $\Delta F = 0$ transition measurements.

[b] Calculated from the experimental hfs interval $F = 7/2 \leftrightarrow F = 5/2$ and the b_{4d}^{02} value obtained from the B factors of the states $^3D_{2,3}$.

48

Table 2. A and B factors for 5d-shell atoms. The level designations and energies are taken from /151,166/

Isotope	Configuration and state	Level energy [cm^{-1}]	A_{exp} [MHz]	δA [kHz]	B_{exp} [MHz]	δB [kHz]	Method	Reference
^{175}Lu	5d6s^2 ^{2}D$_{3/2}$	0	194.330559 (42)	-2.362	1511.39529 (22)	-0.98	ABMR	165
	^{2}D$_{5/2}$	1994	146.777427 (42)	0.955	1860.65354 (58)	-2.59	0	166
	5d^{2}6s ^{4}F$_{3/2}$	18851	-1115.2 (3.0)		285 (30)			
	^{4}F$_{5/2}$	19403	491.7 (4.5)		345 (60)			
	^{4}F$_{7/2}$	20247	764.5 (30.0)		201 (150)			
	^{4}P$_{1/2}$	21472	4904.6 (15.0)					
	^{4}P$_{5/2}$	22802	1540.9 (3.0)		-1088 (9)			
	^{2}D$_{3/2}$	24518	-1064.3 (9.0)		165 (60)			
	^{2}D$_{5/2}$	24711	882.9 (4.5)		-42 (60)			
	^{2}F$_{5/2}$	25861	1115.5 (15.0)		360 (60)			
	^{2}F$_{7/2}$	26570	-440.7 (3.0)		1718 (3)			
	^{2}G$_{7/2}$	27992	-635.6 (15.0)		2422 (60)			
	^{2}P$_{1/2}$	28793	983.3 (6.0)					
	^{2}P$_{3/2}$	29938	-1020.8 (3.0)		-1190 (15)			
	^{2}S$_{1/2}$	30747	7097.6 (3.0)					
^{176}Lu	5d6s^2 ^{2}D$_{3/2}$	0	138.8 (1.0)		2151 (20)		ABMR	167
	^{2}D$_{5/2}$	1994	104.1 (3)		2631 (6)		LIF	168
^{177}Hf	5d^{2}6s ^{3}F$_2$	0	113.43331 (3)	0.17	624.3344 (3)	5.1	ABMR	169
^{179}Hf	5d^{2}6s ^{3}F$_2$	0	-71.42867 (4)	0.24	705.5269 (6)	8.8	ABMR	169
	^{3}F$_3$	2357	-50.8061 (17)	0.0	931.113 (42)	13		170
	^{3}F$_4$	4568	-43.4563 (15)	0.0	1619.128 (55)	28		

Isotope	Config	Term	E (cm⁻¹)	A (MHz)		B (MHz)		Method	Ref
^{181}Ta	$5d^3 6s^2$	$^4F_{3/2}$	0	509.0821 (8)	−1.9	−1012.230 (8)	8	ABMR	171,172
		$^4F_{5/2}$	2010	313.4678 (8)	−1.2	−834.810 (12)	−13		172,173
		$^4F_{7/2}$	3964	264.414 (2)	0	−787.530 (27)	−40		172
		$^4F_{9/2}$	5621	256.617 (2)	0	−650.438 (44)	−50	LIF	174
		$^4P_{1/2}$	6049	884.115 (3)	−53			0	175
		$^4P_{3/2}$	6069	376 (3)		−1370 (26)		LIF	174
	$5d^4 6s$	$^6D_{1/2}$	9759	3109					
		$^6D_{3/2}$	9976	1214 (2)		−18 (12)			
		$^6D_{5/2}$	11244	1220.8 (8)		1060 (11)			
	$5d^3 6s^2$	$^2D_{5/2}$	12866	266.2 (8)		−704 (11)			
^{183}W	$5d^4 6s^2$	$^5D_1$	1670	29.118 (13)	0			ABMR	176
		$^5D_2$	3326	56.261 (8)	0				
		$^5D_3$	4830	78.020 (6)	0				
		$^5D_4$	6219	88.308 (5)	0				
	$5d^5 6s$	$^7S_3$	2951	505.592 (12)	0				
^{185}Re	$5d^5 6s^2$	$^6S_{5/2}$	0	−56.595604 (7)	0.463	29.62365 (8)	−11.56	ABMR	177
		$^4P_{1/2}$	15166	4729 (3)				LIF	66
		$^4P_{3/2}$	13826	1649 (2)		1069 (10)		ABMR	177
		$^4P_{5/2}$	11584	880.4528 (31)	8.5	1618.299 (29)	−230	LIF	66
		$^4G_{5/2}$	14621	2060 (1)		−250 (11)			
		$^4G_{7/2}$	15058	558 (1)		1497 (15)			
		$^4G_{9/2}$	16619	713 (1)		1150 (18)			
		$^4D_{1/2}$	19758	3366 (3)					
		$^4D_{3/2}$	20482	1173 (2)		−527 (7)			
		$^4D_{5/2}$	19458	827 (1)		−264 (10)			
		$^4D_{7/2}$	17331	306.097 (9)	6	329.517 (85)	−451	ABMR	177

Table 2. (cont.)

Isotope	Configuration and state	Level energy $[\text{cm}^{-1}]$	A_{exp} [MHz]	δA [kHz]	B_{exp} [MHz]	δB [kHz]	Method	Reference
	$5d^6 6s$ $\;^6D_{1/2}$	17238	2467.380 (3)	473				
	$^6D_{3/2}$	16328	1456.348 (2)	406	-562.174 (9)	696		
	$^6D_{5/2}$	15770	1072.875 (2)	190	21.130 (15)	3140		
	$^6D_{9/2}$	11755	2584.726 (16)	146	2378.757 (183)	5297		
^{187}Re	$5d^5 6s^2$ $\;^6S_{5/2}$	0	-57.148475 (9)	0.448	28.03384 (9)	-11.76	ABMR	177
	$^4P_{1/2}$	15166	4774 (3)		1019 (10)		LIF	66
	$^4P_{3/2}$	13826	1666 (2)					
	$^4P_{5/2}$	11584	889.2502 (25)	8.9	1531.466 (26)	-235	ABMR	177
	$^4G_{5/2}$	14621	2079 (1)		-240 (11)		LIF	66
	$^4G_{7/2}$	15058	563 (1)		1416 (15)			
	$^4G_{9/2}$	16619	720 (1)		1086 (18)		0	178
	$^4G_{11/2}$	16307	744 (3)		213 (39)		LIF	66
	$^4D_{1/2}$	19758	3399 (3)					
	$^4D_{3/2}$	20482	1186 (1)		-500 (7)			
	$^4D_{5/2}$	19458	834 (1)		-254 (10)			
	$^4D_{7/2}$	17331	309.226 (2)	9	311.784 (29)	-409	ABMR	
	$5d^6 6s$ $\;^6D_{1/2}$	17238	2491.456 (3)	475	-532.035 (11)	704		
	$^6D_{3/2}$	16328	1470.719 (3)	414	20.351 (14)	3238		
	$^6D_{5/2}$	15770	1083.493 (1)	196				
	$^6D_{9/2}$	11755	2610.287 (11)	144	2251.616 (132)	5331		
^{189}Os	$5d^6 6s^2$ $\;^5D_2$	2740	281 (26)		-218 (160)		0	179
	$^5D_3$	4159	119 (14)		198 (90)			

Isotope	Config	Level						Method	Ref
191Ir		5D4	0	170 (10)		1115 (81)			
	5d⁷6s	5F5	5144	948 (6)		390 (45)			
	5d⁷6s²	4F3/2	4097	-41.525565 (25)	-3.255	107.89459 (7)	-2.11	ABMR	180,181
		4F5/2	5785	235.453560 (22)	0.490	337.32949 (10)	-19.51		
		4F7/2	6324	-59.9459 (23)	2.3	-197.169 (17)	-17		
		4F9/2	0	57.521475 (7)	0.195	471.20416 (12)	-0.34		
	5d⁸6s	4F5/2	9878	222.0048 (41)	-0.9	591.385 (22)	45		
		4F7/2	7107	40.6275 (23)	0.5	191.067 (18)	0		
		4F9/2	2835	309.410798 (15)	1.818	-672.34372 (20)	40.78		
193Ir	5d⁷6s²	4F3/2	4097	-44.489911 (21)	-3.511	97.59991 (5)	-1.59	ABMR	180,181
		4F5/2	5785	255.310569 (23)	0.449	305.13480 (10)	-20.40		
		4F7/2	6324	-64.451195 (18)	3.125	-178.39808 (12)	-25.28		
		4F9/2	0	62.655547 (5)	0.177	426.23524 (10)	-0.26		
	5d⁸6s	4F5/2	9878	240.7900 (31)	0.4	534.957 (15)	38		
		4F7/2	7107	44.402514 (22)	0.594	172.87254 (17)	-0.06		
		4F9/2	2835	335.289455 (10)	2.125	-608.16969 (11)	46.51		
195Pt	5d⁸6s²	3F2	15501	1349 (240)			0	ABMR	182
		3F3	10117	1079 (90)					
		3F4	824	849.014 (2)					
		3P1	18567	-150 (300)					
		3P2	6568	1979 (60)					
		1D2	26639	1949 (300)					
		1G4	21967	1109 (90)					
		3D1	10132	-5276 (150)				ABMR	183
		3D2	776	-2607.0355 (35)	2759		0		182
	5d⁹6s	3D3	0	5702.817ᵃ (20)				ABMR	184

Table 2. (cont.)

Isotope	Configuration and state	Level energy [cm^{-1}]	A$_{exp}$ [MHz]		δA [kHz]	B$_{exp}$ [MHz]		δB [kHz]	Method	Reference
197Au	$^1D_2$	13496	5456	(90)	0.0			0.0	0	182
	5d^{9}6s^2 $^2D_{3/2}$	21435	199.8425	(2)		-911.0766	(5)		ABMR	185
	$^2D_{5/2}$	9161	80.236	(3)		-1049.781	(11)			186
	5d^{10}6s $^2S_{1/2}$	0	3049.660092	(7)						163

a Corrected A factor determined from high-field ΔF = 0 transition measurements.

5. Intermediate Coupling Wavefunctions, Atomic g_J Values and Parametric Interpretation of Level Isotope Shifts

In order to evaluate the intermediate coupling wavefunctions the fine-structure parameters were fitted to the experimental level energies as described in Sect.2.5. For the elements Zr, Mo, Rh, Pd, Hf, Ta, Re, Ir, and Pt eigenvectors that span the configurations $(d+s)^N$ were determined in a three-configuration least-squares fit, while for Nb only the two configurations $4d^35s^2$ and $4d^45s$ were taken into account[6]. The same procedure has been applied by other authors to calculate eigenvectors for Tc, Ru, W, and Os (see Tables 3 and 4). In the case of Lu a four-configuration least-squares fit had to be performed due to the overlapping of the $(5d+6s)^3$ configurations with the configuration $6s6p^2$ /187/. For the hfs analysis in the multiplets ds^2 ^2D in Y and Lu and d^9s^2 in Ag and Au pure SL coupling was assumed.

Tables 3 and 4 summarize the fine-structure parameters. Because of the small number of known levels for the 4d- and 5d-shell atoms not all of the radial parameters discussed in Sect.2.5 have been included in the least-squares procedure. Generally, the effective electrostatic two-particle parameter γ and all three-particle effects have been neglected as well as the spin-spin and spin-other-orbit interactions. In order to further reduce the number of free parameters, additional constraints had to be made. In many cases the parameters $F^2(d^N)$, $F^4(d^N)$, and $\zeta(d^N)$, were forced in a linear progression with the parameters of the $d^{N-2}s^2$ and $d^{N-1}s$ configurations by (41) and by the analogous equation for the spin-orbit parameters /87/,

$$\zeta(d^N) - \zeta(d^{N-1}s) = \zeta(d^{N-1}s) - \zeta(d^{N-2}s^2) \quad . \tag{73}$$

Furthermore, if an independent variation of $G^2(d^{N-1}s)$ and $G^2(d^N,d^{N-2}s^2)$ and of $H^2(d^{N-1}s,d^{N-2}s^2)$ and $H^2(d^N,d^{N-1}s)$ led to insignificant values for these parameters, we assumed $G^2(d^{N-1}s) = G^2(d^N,d^{N-2}s^2)$ and $H^2(d^{N-1}s,d^{N-2}s^2) = H^2(d^N,d^{N-1}s)$.

---

[6] The basic version of the computer program was kindly supplied by W.J. Childs, Argonne National Laboratory.

Table 3. Fine-structure parameters (in cm^{-1}) for 4d-shell atoms

Element	Y	Zr	Nb	Mo	Tc	Ru	Rh	Pd	Ag
$F^2(d^{N-2}s^2)$		32396 (570)	37925 (500)	43712 (817)	45249 (666)	48421 (233)	51422 (90)	55517	
$F^2(d^{N-1}s)$		27621 (315)	31649 (254)	36335 (182)	40737 (525)	44638 (166)	48263 (170)		
$F^2(d^N)$		22847		25316 (882)	36223	40688 (542)			
$F^4(d^{N-2}s^2)$		18025 (906)	25572 (434)	25611 (286)	30610 (353)	34529 (227)	37160 (224)	42336	
$F^4(d^{N-1}s)$		15339 (411)	18921 (272)	24189 (147)	29111 (340)	32490 (139)	35097 (472)		
$F^4(d^N)$		12653		24100	27610	30399 (756)			
$G^2(d^{N-1}s)$		9021 (56)	8551 (71)	9116 (68)	8820 (75)	8680 (55)	8496 (51)	7908 (1025)	
$\alpha(d^{N-2}s^2)$		27 (7)	-7 (7)	27 (2)	28 (3)	29 (1)	37 (5)	28	
$\alpha(d^{N-1}s)=\alpha(d^N)$		28 (3)	50 (4)	27	28	29	37		
β	212	185 (160)		190 (56)					
$\zeta(d^{N-2}s^2)$		344 (18)	515 (35)	685 (32)	744 (72)	971 (23)	1229 (7)	1487 (22)	1789
$\zeta(d^{N-1}s)$		298 (14)	393 (24)	560 (26)	674 (26)	916 (16)	1159 (13)	1403 (17)	
$\zeta(d^N)$		257 (25)		422 (60)	605	838 (30)	1081 (20)		
$G^2(d^N,d^{N-2}s^2)$		9021		11093 (500)	8820	8680	8496	7908	
$H^2(d^{N-1}s,d^{N-2}s^2)$		-12262 (116)	-14406 (202)	-12054 (122)	-11774 (140)	-11547 (70)	-11138 (113)	-9261 (4845)	
$H^2(d^N,d^{N-1}s)$		-12262		-12074 (204)	-11774	-11547	-11678 (320)		
Q^2				65 (41)					
Q^4				45					
Number of levels fitted	2	39	49	78	48	53	26	8	
$\sigma(cm^{-1})$		55	109	89	107	63	21	34	
Reference		153	188	189	158	190	191	192	2

Table 4. Fine-structure parameters (in cm^{-1}) for 5d-shell atoms

Element	Lu	Hf	Ta	W	Re	Os	Ir	Pt	Au
$F^2(d^{N-2}s^2)$		29917 (377)	32515 (316)	37121 (373)	40389 (366)	43108 (343)	45675 (517)	49022 (82)	
$F^2(d^{N-1}s)$	21427 (946)	25113 (341)	30062 (448)	34048 (370)	36774 (585)	41026 (392)	43230 (537)		
$F^2(d^N)$	18976	20310	27607	30975	33193	38945			
$F^4(d^{N-2}s^2)$		15121 (557)	18421 (363)	23688 (265)	27560 (451)	32361 (617)	32754 (760)	37331 (181)	
$F^4(d^{N-1}s)$	9463(1071)	11684 (501)	16965 (360)	21508 (252)	26098 (389)	31218 (573)	32739 (978)		
$F^4(d^N)$	7592	8251	15523	19328	24634	30076			
$G^2(d^{N-1}s)$	10740 (295)	12424 (162)	12963 (262)	13715 (120)	13592 (111)	13557 (150)	13355 (222)	13558 (83)	
$\alpha(d^{N-2}s^2)$		40 (10)	44 (6)	39 (2)	24 (4)	16 (7)	21 (6)	25 (2)	
$\alpha(d^{N-1}s)=\alpha(d^N)$		52 (4)	26 (5)	39	24	15 (7)	21		
β		130 (138)	250 (53)		192 (162)	-262 (140)[a]	368 (211)		
$\zeta(d^{N-2}s^2)$	812 (99)	1211 (30)	1693 (20)	2104 (20)	2560 (19)	3058 (27)	3574 (21)	4182 (14)	4910
$\zeta(d^{N-1}s)$	674 (55)	1084 (19)	1428 (23)	1888 (34)	2290 (20)	2836 (28)	3367 (41)	3964 (23)	
$\zeta(d^N)$	535	801 (40)	1163	1672	2020	2613	3210 (143)		
$G^2(d^N,d^{N-2}s^2)$	10740	12424	12963	13715	12288(1019)	13557	15781 (967)	13558	
$H^2(d^{N-1}s,d^{N-2}s^2)$	-12250	-14647 (354)	-15126 (608)	-16625 (210)	-18593 (227)	-18816 (175)	-19126 (131)	-19148 (110)	
$H^2(d^N,d^{N-1}s)$	-11550	-12351 (601)	-12665(1371)	-16625	-15793 (653)	-18816	-17610 (539)		
Q^2					284 (71)		383 (125)	253 (67)	
Q^4					230 (207)		268	177	
Number of levels fitted	19	47	41	57	55	49	31	12	2
$\sigma(\text{cm}^{-1})$	181	106	79	100	63	110	68	15	
Reference	187	192	172	193	177	194	181	184	

[a] For Os the effective electrostatic interaction has been written as βQ instead of $\beta G(R_5)$, where Q is the seniority operator /76/.

The number of independent parameters Q^k of the electrostatically correlated spin-orbit interaction was reduced by $Q^k(d^{N-2}s^2) = Q^k(d^{N-1}s) = Q^k(d^N)$, $k = 2,4$, and in some cases we used the estimate $Q^4 = 0.7 \, Q^2$ found from Hartree-Fock calculations for 3d-shell atoms /195/. Which parameters have been included in the least-squares procedure in each individual element can be seen from Tables 3 and 4. The rms error, which is defined as

$$\sigma = \sqrt{\frac{\sum\limits_{i=1}^{n} (E^i_{calc} - E^i_{exp})^2}{(n-m)}} \tag{74}$$

where n is the number of fitted levels and m is the number of free parameters, is smaller than 110 cm^{-1} in all cases except for Lu.

In order to test the eigenvectors evaluated in this way, the g_J values have been calculated according to (53). In Table 5 these theoretical values, which include the Schwinger correction ($g_s = 2.002319$), but neglect relativistic and diamagnetic corrections, are compared with experimental g_J values obtained with high precision by ABMR. The agreement found is very good in most cases, typically better than 0.1% for 4d-shell atoms and 1% for 5d-shell atoms. The residual differences between experimental and theoretical g_J values are due to relativistic and diamagnetic corrections, which are expected to be of the order of some parts in 10^{-4}, as well as to slight inaccuracies in the intermediate coupling eigenvectors. Apart from these deficiencies, which are caused by the approximations which had to be made because of the lack of experimental level energies, the quality of the eigenvectors can be demonstrated by comparing the differences between experimental and theoretical g_J factors of strongly perturbed levels with the departures of the observed g_J values from the SL limit. For example, using (52) one calculates for the states $4d^2 5s^2 \; {}^3P_2$ and ${}^1D_2$ in Zr $(g_J^{exp} - g_J^{SL})({}^3P_2) = -0.23644$ and $(g_J^{exp} - g_J^{SL})({}^1D_2) = 0.23146$, while the differences between the experimental g_J values and those obtained in intermediate coupling are -0.00010 and 0.00002, respectively. Another example are the g_J factors of the $5d^6 6s \; {}^6D_{1/2,3/2,5/2,9/2}$ states in Re. The departures from the SL limit are -0.80502, -0.15737, -0.34095, and -0.02220, respectively. As can be seen from Table 5 the differences between observed and intermediate coupling g_J values are smaller by a factor 40 to 100.

From the atomic beam laser fluorescence measurements, performed in connection with the hfs investigation of high-lying metastable states by ABMR with laser detection (Sect.3.3), one obtains not only rough values for the hfs constants but also the IS of the spectral lines used for excitation from the metastable states. Choosing one level as a reference level, one may derive from such measurements the IS of the levels with respect to the reference level. In the case of Re, for exam-

Table 5. Comparison between experimental and calculated g_J factors. The level designations are taken from /151,166/

Element	Configuration and state	g_J^{exp}	Reference	g_J^{calc}	$g_J^{exp} - g_J^{calc}$	Largest eigenvector component %
Y	$4d5s^2$ $^2D_{3/2}$	0.79927 (11)	196	0.79954	-0.00027	$4d5s^2$ ^2D 100
	$^2D_{5/2}$	1.20028 (19)		1.20046	-0.00018	$4d5s^2$ ^2D 100
Zr	$4d^25s^2$ $^3F_2$	0.66981 (4)	197	0.66994	-0.00013	$4d^25s^2$ ^3F 96
	$^3F_3$	1.08331 (9)		1.08352	-0.00021	$4d^25s^2$ ^3F 97
	$^3F_4$	1.24987 (6)		1.25009	-0.00022	$4d^25s^2$ ^3F 96
	$^3P_1$	1.50072 (12)		1.50106	-0.00034	$4d^25s^2$ ^3P 73
	$^3P_2$	1.26472 (5)		1.26482	-0.00010	$4d^25s^2$ ^1D 40
	$^1D_2$	1.23146 (3)		1.23144	0.00002	$4d^25s^2$ ^1D 45
	$^1G_4$	1.00052 (5)		1.00072	-0.00020	$4d^25s^2$ ^1G 81
	$4d^35s$ $^5F_2$	1.00081 (10)		1.00153	-0.00072	$4d^35s$ ^5F 99
	$^5F_3$	1.25021 (5)		1.25038	-0.00017	$4d^35s$ ^5F 100
	$^5F_4$	1.35012 (4)		1.35024	-0.00012	$4d^35s$ ^5F 100
	$^5F_5$	1.39991 (4)		1.39999	-0.00008	$4d^35s$ ^5F 100
Nb	$4d^35s^2$ $^4F_{3/2}$	0.40343 (15)	154	0.40406	-0.00063	$4d^35s^2$ ^4F 92
	$^4F_{5/2}$	1.03043 (4)		1.03124	-0.00081	$4d^35s^2$ ^4F 92
	$^4F_{7/2}$	1.23914 (3)		1.23990	-0.00076	$4d^35s^2$ ^4F 90
	$^4F_{9/2}$	1.33363 (3)		1.33426	-0.00063	$4d^35s^2$ ^4F 88
	$4d^45s$ $^6D_{1/2}$	3.32901 (7)		3.32845	0.00056	$4d^45s$ ^6D 98
	$^6D_{3/2}$	1.86630 (3)		1.86607	0.00023	$4d^45s$ ^6D 98
	$^6D_{5/2}$	1.65640 (3)		1.65609	0.00031	$4d^45s$ ^6D 100
	$^6D_{7/2}$	1.58583 (3)		1.58550	0.00033	$4d^45s$ ^6D 100
	$^6D_{9/2}$	1.55328 (2)		1.55298	0.00030	$4d^45s$ ^6D 98

Table 5. (cont.)

Element	Configuration and state	g_J^{exp}	Reference	g_J^{calc}	$g_J^{exp} - g_J^{calc}$	Largest eigenvector component %
Mo	$4d^45s$ $^7S_3$	2.00055 (4)	140	2.00082	-0.00027	$4d^55s$ ^7S 100
Ru	$4d^75s$ $^5F_2$	1.00112 (1)	159	1.001	0.000	$4d^75s$ ^5F 99
	$^5F_3$	1.24899 (2)		1.249	0.000	$4d^75s$ ^5F 99
	$^5F_4$	1.34760 (2)		1.347	0.001	$4d^75s$ ^5F 98
	$^5F_5$	1.39774 (2)		1.397	0.001	$4d^75s$ ^5F 99
Rh	$4d^85s$ $^4F_{3/2}$	0.47399 (2)	161	0.47650	-0.00251	$4d^85s$ ^4F 81
	$^4F_{5/2}$	1.10835 (3)		1.11058	-0.00223	$4d^85s$ ^4F 53
	$^4F_{7/2}$	1.23144 (4)		1.23127	0.00017	$4d^85s$ ^4F 93
	$^4F_{9/2}$	1.33253 (3)		1.33297	-0.00044	$4d^85s$ ^4F 99
	$^2F_{5/2}$	0.94765 (4)		0.95508	-0.00743	$4d^85s$ ^2F 78
	$^2F_{7/2}$	1.14710 (4)		1.14819	-0.00109	$4d^85s$ ^2F 92
	$4d^9$ $^2D_{3/2}$	0.74414 (4)		0.74295	0.00119	$4d^9$ ^2D 71
	$^2D_{5/2}$	1.11818 (3)		1.11631	0.00187	$4d^9$ ^2D 49
Pd	$4d^95s$ $^3D_1$	0.49809 (2)	162	0.49884	-0.00075	$4d^95s$ ^3D 100
	$^3D_2$	1.12747 (2)		1.12712	0.00035	$4d^95s$ ^3D 76
	$^3D_3$	1.33368 (2)		1.33411	-0.00043	$4d^95s$ ^3D 100
	$^1D_2$	1.03856 (2)		1.03991	-0.00135	$4d^95s$ ^1D 75
Ag	$4d^{10}5s$ $^2S_{1/2}$	2.002347 (4)	163	2.00232	0.00003	$4d^{10}5s$ ^2S 100
Lu	$5d6s^2$ $^2D_{3/2}$	0.79921 (8)	198	0.79954	-0.00033	$5d6s^2$ ^2D 100
	$^2D_{5/2}$	1.20040 (16)		1.20046	-0.00006	$5d6s^2$ ^2D 100
Hf	$5d^26s^2$ $^3F_2$	0.695812 (10)	199	0.69635	-0.00054	$5d^26s^2$ ^3F 90
	$^3F_3$	1.083367 (2)	170	1.08352	-0.00015	$5d^26s^2$ ^3F 98

58

Element	Config	Term	Value (unc.)		Value	Correction	Config	No.
Ta	$5d^36s^2$	$^3F_4$	1.240743	(7)		1.24132	-0.00058	$5d^26s^2\ ^3F$	94
		$^4F_{3/2}$	0.45024	(4)	171	0.45842	-0.00818	$5d^36s^2\ ^4F$	85
		$^4F_{5/2}$	1.03499	(4)	172	1.03625	-0.00126	$5d^36s^2\ ^4F$	92
		$^4F_{7/2}$	1.21823	(4)		1.21623	0.00200	$5d^36s^2\ ^4F$	90
		$^4F_{9/2}$	1.28760	(3)		1.28302	0.00458	$5d^36s^2\ ^4F$	76
		$^4P_{1/2}$	2.46921	(29)	176	2.45964	0.00957	$5d^36s^2\ ^4P$	64
W	$5d^46s^2$	$^5D_1$	1.49926	(7)		1.498	0.001	$5d^46s^2\ ^5D$	84
		$^5D_2$	1.48683	(8)		1.485	0.002	$5d^46s^2\ ^5D$	88
		$^5D_3$	1.47813	(8)		1.475	0.003	$5d^46s^2\ ^5D$	85
		$^5D_4$	1.45514	(7)		1.452	0.003	$5d^46s^2\ ^5D$	77
Re	$5d^56s$	$^7S_3$	1.98142	(5)	177	1.978	0.003	$5d^56s\ ^7S$	94
	$5d^56s^2$	$^6S_{5/2}$	1.95199	(5)		1.95080	0.00119	$5d^56s^2\ ^6S$	88
		$^4P_{5/2}$	1.27944	(2)		1.28105	-0.00161	$5d^56s^2\ ^4D$	19
		$^4G_{5/2}$	1.14948	(15)		1.15424	-0.00476	$5d^56s^2\ ^4G$	36
		$^4D_{7/2}$	1.25465	(17)	66	1.25691	-0.00226	$5d^56s^2\ ^4D$	39
	$5d^66s$	$^6D_{1/2}$	2.53372	(17)		2.55611	-0.02239	$5d^66s\ ^6D$	61
		$^6D_{3/2}$	1.71131	(7)		1.71443	-0.00312	$5d^66s\ ^6D$	75
		$^6D_{5/2}$	1.31772	(3)		1.31428	0.00344	$5d^66s\ ^6D$	49
		$^6D_{9/2}$	1.53464	(5)	180	1.53525	-0.00061	$5d^66s\ ^6D$	91
Ir	$5d^76s^2$	$^4F_{3/2}$	1.12311	(5)		1.11653	0.00658	$5d^76s^2\ ^2P$	34
		$^4F_{5/2}$	1.19912	(5)		1.19666	0.00246	$5d^76s^2\ ^4F$	23
	$5d^86s$	$^4F_{7/2}$	1.20121	(5)		1.19882	0.00239	$5d^86s\ ^4F$	57
		$^4F_{9/2}$	1.29694	(3)	200	1.29717	-0.00023	$5d^76s^2\ ^4F$	84
		$^4F_{5/2}$	1.17109	(5)	180	1.17955	-0.00846	$5d^76s^2\ ^4F$	32
		$^4F_{7/2}$	1.22464	(5)		1.22511	-0.00047	$5d^76s^2\ ^4F$	82
		$^4F_{9/2}$	1.32748	(4)		1.32759	-0.00011	$5d^86s\ ^4F$	96

Table 5. (cont.)

Element	Configuration and state	g_J^{exp}	Reference	g_J^{calc}	$g_J^{exp} - g_J^{calc}$	Largest eigenvector component %
Pt	$5d^8 6s^2\ ^3P_2$	1.11691 (3)	201	1.11815	-0.00124	$5d^9 6s\ ^3D$ 42
	$^3F_4$	1.23948 (2)		1.24000	-0.00052	$5d^8 6s^2\ ^3F$ 96
	$5d^9 6s\ ^3D_2$	1.066574 (3)	184	1.06795	-0.00138	$5d^9 6s\ ^1D$ 41
	$^3D_3$	1.333924 (3)		1.33411	-0.00019	$5d^9 6s\ ^3D$ 100
Au	$5d^9 6s^2\ ^2D_{3/2}$	0.799 (1)	185	0.79954	-0.001	$5d^9 6s^2\ ^2D$ 100
	$^2D_{5/2}$	1.20000 (2)	186	1.20046	-0.00046	$5d^9 6s^2\ ^2D$ 100
	$5d^{10} 6s\ ^2S_{1/2}$	2.003312 (4)	163	2.00232	0.00099	$5d^{10} 6s\ ^2S$ 100

Table 6. Values of IS parameters for 5d-shell atoms (in MHz)

Isotope pair	s	g_2	h	z_s^2	Reference
$184,186$W	-1889 (90)	0^a	57 (9)	51 (27)	204
$185,187$Re	-4253 (463)	-308 (93)	34 (3)	27 (13)	66
$190,192$Os	-2638 (150)	-420 (90)	69 (18)	0^a	202,204
$194,196$Pt	-3508 (120)	-450 (180)	42 (21)	60 (63)	204

a Fixed

ple, the IS of 13 levels of the configurations $(5d+6s)^7$ could be determined from the laser fluorescence measurements /66/ combined with results of conventional optical spectroscopy /178/.

Considerable progress has been achieved, during recent years, in the parametric description of level IS, which is based on the central field model /202/. Using the intermediate coupling wavefunctions the IS of each level may be written as a numerical expansion of IS parameters. These parameters are determined by a least-squares fit procedure, so that the experimental IS is reproduced in the most accurate way. An advantage of this method is that the field shift and the mass shift need not necessarily be separated. The full accuracy of the measurements is therefore available for the determination of the IS parameters. This parametric method takes into account the first-order contributions from the normal mass shift, the specific mass shift, and the field shift operator as well as the influence of relativistic and CI effects. Thus, the parametric analysis of level IS may provide valuable information on relativistic and CI effects in the IS. A detailed discussion of the phenomenological description of IS has recently been given by BAUCHE and CHAMPEAU /203/. Table 6 summarizes the values of IS parameters found for some 5d elements from conventional optical spectroscopy (W, Os, Pt) and from conventional optical spectroscopy combined with laser fluorescence spectroscopy (Re). The parameter $s = \Delta T(5d^{N-1}6s) - \Delta T(5d^{N-2}6s^2)$ describes the difference between the contributions ΔT which are constant in the configurations. g_2 and h, with the same angular coefficients as the Slater integrals G^2 and H^2, and z_{s2}, with the same angular coefficients as the spin-orbit constant $\zeta(5d^{N-2}6s^2)$, take into account relativistic and second-order CI effects. The analogies in the IS behavior in the four 5d-shell atoms W, Re, Os, and Pt are remarkable; however, the rms errors are still rather large due to the experimental uncertainties in conventional optical spectroscopy. This may be considerably improved by purely laser spectroscopic measurements of IS in the near future.

6. Effective Radial Parameters of the Magnetic Dipole and Electric Quadrupole Interaction

6.1 Determination of Effective Radial Parameters from Experiments

Using the intermediate coupling wavefunctions the hfs constants A and B of the magnetic dipole and electric quadrupole interaction were expressed as linear combinations of ten $a^{k_s k_l}$ and nine $b^{k_s k_l}$ parameters, respectively, as described in Sect. 2.5[7]. These expressions are, however, not immediately useful in determinating the $a^{k_s k_l}$ and $b^{k_s k_l}$ parameters from the experimental hfs constants since for the bulk of the 4d- and 5d-shell atoms the number of investigated metastable states is too small. Furthermore, except for the $4d^9\,{}^2D$ multiplet in Rh, no A's and B's were measured in states with appreciable admixtures from d^N configurations. Therefore, the parameters for the d^N configurations either were fixed at estimated values or were expressed through the corresponding parameters of the $d^{N-2}s^2$ configurations by the relations

$$a^{k_s k_l}(d^N) = \frac{\zeta(d^N)}{\zeta(d^{N-2}s^2)}\, a^{k_s k_l}(d^{N-2}s^2) \tag{75a}$$

$$b^{k_s k_l}(d^N) = \frac{\zeta(d^N)}{\zeta(d^{N-2}s^2)}\, b^{k_s k_l}(d^{N-2}s^2) \tag{75b}$$

because the spin-orbit constants are like the $a^{k_s k_l}$ and $b^{k_s k_l}$ parameters approximately proportional to $<r^{-3}>$.

In several cases the relation

$$a_d^{10}(d^N) = a_d^{10}(d^{N-2}s^2) \tag{75c}$$

was used for the d-electron contact parameter. The errors introduced by these approximations can be neglected because of the small coefficients of the d^N parame-

---

[7] The coefficients of these expressions were evaluated using the computer program ABCFAK written by S. Büttgenbach, M. Herschel, and F. Träber, University of Bonn.

ters in the numerical expressions for the A and B factors of the investigated levels. Thus, the number of effective radial parameters is reduced to seven a's and six b's. For some 4d- and 5d-shell atoms all of these parameters could be determined; however, in most cases further constraints had to be made because of the small number of measured hfs constants or because of mathematical correlations in the system of linear equations which led to insignificant parameter values. In Tables 7 and 8 the adjusted parameter values are summarized. No errors are quoted for the a's and b's since the experimental uncertainties can be neglected compared to errors which are introduced by the approximations made at all stages of the analysis of the experimental data and which are hard to estimate.

Some pecularities concerning the individual isotopes shall be discussed in the following.

^{89}Y. Assuming pure SL coupling one finds from the A factors of the $4d5s^2$ ^2D multiplet $(a^{01} + \frac{1}{7} a^{12}) = -43.0$ MHz. Because the parameter a^{01} generally turns out to be less sensitive to CI effects than a^{12} or a^{10} (Sect.6.2), $a^{01}(4d5s^2)$ was fixed at the MCDF value. Then, $a^{12}(4d5s^2)$ and $a^{10}(4d5s^2)$ were evaluated from the A factors of the ^2D multiplet.

^{91}Zr. Again, $a^{01}(4d^2 5s^2)$ was fixed at the MCDF value; the remaining six a's were adjusted to seven experimental A factors. The mean deviation of the recalculated A values from the experimental ones is 0.1%. For the adjustment of six b's to seven experimental B factors the mean deviation is 0.1%, too.

^{93}Nb. Seven a and six b parameters were determined from nine A and eight B factors. The mean deviations are 0.02% and 0.5%, respectively.

^{97}Mo. Using the ratio of the nuclear magnetic dipole moments the A factor of the $^5S_2$ state was evaluated from the experimental value of $A(^{95}Mo, ^5S_2)$. $a^{01}(4d^4 5s^2)$ was fixed at the MCDF value, and the parameters $a^{01}(4d^5 5s)$ and $a^{12}(4d^5 5s)$ were related to the corresponding $4d^4 5s^2$ parameters by the ratio of the spin-orbit constants. The remaining four a's were fitted to six A factors with a mean deviation of 4.1%. The b parameters of the configurations $4d^4 5s^2$ and $4d^5 5s$ were related by the ratio of the spin-orbit constants, too. Then, three b's were fitted to five B factors with a mean deviation of 15%.

^{99}Tc. The parameter values in Table 7 are taken from /158/. $a^{01}(4d^6 5s)$ and $a^{12}(4d^6 5s)$ were related to the corresponding parameters of the configuration $4d^5 5s^2$ by the ratio of the spin-orbit constants. The remaining five a's were adjusted to the experimental

Table 7. Effective radial parameters of the magnetic dipole and electric quadrupole interaction in 4d-shell atoms

Isotope	I	μ_I^a [μ_N]	Configuration	a^{01} [MHz]	a^{12} [MHz]	a^{10}_{4d} [MHz]	a^{10}_{5s} [MHz]	b^{02} [MHz]	b^{13} [MHz]	b^{11} [MHz]
89Y	1/2	-0.13686	$4d5s^2$	-42.0[b]	-7.1	20.1				
91Zr	5/2	-1.29806	$4d^25s$	-114.2[b]	-71.7	42.3		-119.3	-14.8	3.3
93Nb	9/2	6.1431	$4d^35s$	-95.4	-54.0	68.1	-1906	-87.8	-3.2	13.8
			$4d^35s^2$	395.4	238.3	-225.8		-263.6	-30.6	7.3
97Mo	5/2	-0.9292	$4d^45s$	351.7	276.1	-511.5	6700	-232.0	-27.5	2.6
			$4d^45s^2$	-134.8[b]	-81.4	71.5		229.2	55.3	-7.9
99Tc	9/2	5.6575	$4d^55s$	-110.3[c]	-66.6[c]	114.0	-1847	187.5[c]	45.3[c]	-6.5[c]
			$4d^55s^2$	533.6[c]	533.6[c]	-109.1		-129[f,d]		
101Ru	5/2	-0.7152	$4d^65s$	482.7	482.7	-314.8	6580	-129[d]		
103Rh	1/2	-0.08794	$4d^75s$	-134.4	-104.1	63.6	-1662	532.1	52.4	-70.3
105Pd	5/2	-0.639	$4d^85s$	-100.0	-66.8	12.8	-1007			
109Ag	1/2	-0.12984	$4d^95s$	-180.5	-180.5[e]	65.0	-1794	1133[d]		
			$4d^95s^2$	-223.3	-223.3[e]	39.7				
			$4d^{10}5s$				-1977			

a The magnetic dipole moments (uncorrected for diamagnetism) are taken from /1/.

b The parameter has been fixed at the quoted value.

c The ratio of the parameters for the configurations $d^{N-1}s$ and $d^{N-2}s^2$ has been fixed at the ratio of the spin-orbit constants.

d $b^{13} = b^{11} = 0$ has been assumed.

e $a^{12} = a^{01}$ has been assumed.

f $b^{02}(d^{N-2}s^2) = b^{02}(d^{N-1}s)$ has been assumed.

Table 8. Effective radial parameters of the magnetic dipole and electric quadrupole interaction in 5d-shell atoms

Isotope	I	μ_I^a [μ_N]	Configuration	a^{01} [MHz]	a^{12} [MHz]	a^{10}_{5d} [MHz]	a^{10}_{6s} [MHz]	b^{02} [MHz]	b^{13} [MHz]	b^{11} [MHz]
^{175}Lu	7/2	2.21102	$5d6s^2$	181.8[b]	-78.6	-38.1		3511	1265	-431[c]
^{179}Hf	9/2	-0.6329	$5d^26s^2$	158.0	158.0[d]	-449.7	8166	4017	11	0[b]
^{181}Ta	7/2	2.341	$5d^26s^2$	-54.9	-6.1	1.8		5269	2627	-997
			$5d^36s^2$	324.6	94.0	-56.1	5247	4839	2217	-211
^{183}W	1/2	0.11621	$5d^46s$	376.9[e]	79.3[e]	884.0		4895	1870[e]	-178[e]
			$5d^46s^2$	145[b]	75	-61				
			$5d^56s$			-61[f]	3315			
^{187}Re	5/2	3.1761	$5d^56s^2$	836.7	958.0	-150.0		4585	2686	-297
			$5d^66s$	777.4	856.9[e]	185.0	18326	3999	2403[e]	-265[e]
			$d^5s^2-d^6s$		1394			-457		
^{189}Os	3/2	0.65063	$5d^66s^2$	374.2[b]	138.4	-124.7		2085	1078	-504
			$5d^76s$	347.0[e]	128.4[e]	-124.7[f]	7725	1518	1000[e]	-467[e]
^{193}Ir	3/2	0.1568	$5d^76s^2$	137.4	180.8	-101.8		2115	854	-204
			$5d^86s$	129.4[e]	170.3[e]	-95.9[e]	2547	1992[e]	804[e]	-192[e]
^{195}Pt	1/2	0.60039	$5d^86s^2$	1331.0	515.5	-262.7				
			$5d^96s$	1261.8[e]	488.7[e]	-262.7[f]	29408			
^{197}Au	3/2	0.14349	$5d^96s^2$	127.0	91.3[b]	-54.6	3050	1985	1056	-349[c]
			$5d^{10}6s$							

65

a The magnetic dipole moments (uncorrected for diamagnetism) are taken from /1/.
b The parameter has been fixed at the quoted value.
c The ratio b^{11}/b^{13} has been fixed at the MCDF value.
d $a^{12} = a^{01}$ has been assumed.
e The ratio of the parameters for the configurations $d^{N-1}s$ and $d^{N-2}s^2$ has been fixed at the ratio of the spin-orbit constants.
f $a^{10}_{5d}(d^{N-1}s) = a^{10}_{5d}(d^{N-2}s^2)$ has been assumed.

A factors. The mean deviation between experimental and recalculated A's is 4.4%. For the electric quadrupole interaction $b^{11} = b^{13} = 0$ and $b^{02}(4d^6 5s) = b^{02}(4d^5 5s^2)$ was assumed. The result is $b^{02} = -129$ MHz, which reproduces the experimental B factors within their large errors bars.

[101]Ru. Since the hfs had been measured in four states only, the radial parameters of the configuration $4d^6 5s^2$, whose contributions to the hfs constants of the $4d^7 5s\ ^5F$ multiplet are very small, were fixed in the following way: $a^{01}(4d^6 5s^2) = a^{12}(4d^6 5s^2) = a^{01}(4d^7 5s)$, $b^{02}(4d^6 5s^2) = b^{02}(4d^7 5s)$, and $a^{10}(4d^6 5s^2) = b^{13}(4d^6 5s^2) = b^{11}(4d^6 5s^2) = 0$. The remaining four a's and three b's were evaluated from the four A and B factors, respectively. The mean deviation of the recalculated B factors from the experimental results is 0.6%.

[103]Rh. The four a's of the configuration $4d^8 5s$ have been varied independently to fit the eight experimental A factors, while the a's of the configuration $4d^7 5s^2$ and $4d^9$ were related to the a's of the configuration $4d^8 5s$ using the respective ratios of the spin-orbit constants. The mean deviation of the recalculated A values from the experimental results is 8%.

[105]Pd. Since the expressions for the magnetic dipole constants of the $4d^9 5s\ ^3D$ and ^1D multiplets turned out to be linearly dependent, $a^{01} = a^{12}$ has been assumed. The remaining three a's were fitted to four A factors with a mean deviation of 2.9%. For the electric quadrupole interaction $b^{13} = b^{11} = 0$ has been assumed. The resulting value for b^{02} reproduces the experimental B factors with a mean deviation of 6%.

[109]Ag. From the ground-state hyperfine structure one immediately finds the value for a_{5s}^{10}. For the $4d^9 5s^2\ ^2D$ multiplet $a^{01} = a^{12}$ has been assumed.

[175]Lu. Assuming pure SL coupling and fixing a^{01} at the MCDF value one finds $a^{12}(5d6s^2)$ and $a_{5d}^{10}(5d6s^2)$ from the A factors of the ^2D ground multiplet. For the B factors the ratio b^{13}/b^{11} has been fixed at the MCDF value. For the configuration $5d^2 6s$ the a's and b's determined by WYART /187/ are quoted in Table 8.

[179]Hf. Since only three states of the configuration $5d^2 6s^2$ have been investigated experimentally, the b parameters have been evaluated from the B factors neglecting the contributions from the configurations $5d^3 6s$ and $5d^4$, which are smaller than 1%. The parameters $a^{01}(5d^3 6s)$ and $a^{12}(5d^3 6s)$ were related to the corresponding parameters of the configuration $5d^2 6s^2$ using the ratio of the spin-orbit constants. The contact parameters a_{5d}^{10} have been assumed to be equal for the two configurations.

For the contact parameter of the 6s-electron the value a_{6s}^{10} = -1275 MHz has been estimated from the results for the neighboring elements. Because of the scattering of the $<r^{-3}>_{6s}^{10}$ values in the region Lu - W (Fig.11b) a large uncertainty has to be assigned to this estimate; a_{5d}^{10} and a^{12} depend strongly on the choice of a_{6s}^{10}, while a^{01} is fairly insensitive.

^{181}Ta. The ratio $a^{12}(5d^46s)/a^{12}(5d^36s^2)$ has been fixed at the ratio of the respective spin-orbit constants. The remaining six a's were fitted to ten A factors. The mean deviation between recalculated A's and experimental results is 4.2%. The ratio of the b^{13} and b^{11} parameters for the two configurations $5d^46s$ and $5d^36s^2$ has also been expressed by the ratio of the spin-orbit constants, and four b's were fitted to eight experimental B factors with a mean deviation of 10.4%.

^{183}W. In the SL limit the parameters $a_{5d}^{10}(5d^56s)$ and $a_{6s}^{10}(5d^56s)$ are linearly dependent, as are, because of S = L for the ^{5}D multiplet, the parameters $a^{01}(5d^46s^2)$ and $a_{5d}^{10}(5d^46s^2)$. The existence of such correlations led to irregular values for the radial parameters of the $5d^46s^2$ configuration, even in an atom so far away from pure SL coupling as W. In order to reduce the number of free parameters, the parameters $a^{01}(5d^56s)$ and $a^{12}(5d^56s)$ were related to the corresponding parameters of the configuration $5d^46s^2$ using the ratio of the spin-orbit constants, and $a_{5d}^{10}(5d^56s)$ was held equal to $a_{5d}^{10}(5d^46s^2)$. Furthermore, $a^{01}(5d^46s^2)$ was fixed at its MCDF value. The remaining three a's were fitted to five experimental A factors with a mean deviation of 2.5%.

^{187}Re. The ratio of the parameters a^{12}, b^{13}, and b^{11} for the configurations $5d^66s$ and $5d^56s^2$ has been fixed at the ratio of the respective spin-orbit constants. Since sixteen A's and thirteen B's had been determined experimentally, the a and b parameters between the configurations were also determined assuming $a^{12}(5d^66s,5d^56s^2)$ = $a^{12}(5d^7,5d^66s)$ and $b^{02}(5d^66s,5d^56s^2)$ = $b^{02}(5d^7,5d^66s)$. Thus, a total of seven a and five b parameters were determined with a mean deviation of 7.4% for the A factors and 21% for the B factors (neglecting the ^{6}D$_{5/2}$ state, for which the small experimental B factor of 17 MHz cannot be reproduced well). From Table 8 it can be seen that the off-diagonal a^{12} parameter is of the same order of magnitude as the diagonal a^{12} parameters.

^{189}Os. In order to reduce the number of free a parameters the ratios of a^{01} and a^{12} for the configurations $5d^66s^2$ and $5d^76s$ were again expressed by the ratio of the spin-orbit constants. The contact parameters a_{5d}^{10} were assumed to be equal for the two configurations. Furthermore, $a^{01}(5d^66s^2)$ was fixed at its MCDF value. The remaining a parameters, namely $a^{12}(5d^6 6s^2)$, $a^{10}_{5d}(5d^6 6s^2)$, and $a^{10}_{6s}(5d^7 6s)$ were fitted to four experimental A factors. The mean deviation between recalculated and experimental A factors is 4.9%. The b parameters for the configuration $5d^6 6s^2$ were determined from the B factors of the $5d^6 6s^2$ $^5D_{2,3,4}$ states. Then, $b^{02}(5d^7 6s)$ was evaluated from the B factor of the $5d^7 6s$ $^5F_5$ state. For this purpose $b^{13}(5d^7 6s)$ and $b^{11}(5d^7 6s)$ were related to the corresponding parameters for the configuration $5d^6 6s^2$ using the ratio of the spin-orbit constants.

^{193}Ir. In order to obtain physically realistic parameter values the parameters of the $5d^8 6s$ configuration were expressed by the parameters of the $5d^7 6s^2$ configuration by the ratio of the spin-orbit constants. The remaining four a's and three b's were fitted to seven A's and six B's, respectively. The mean deviations were 17% for the A factors and 32% for the B factors.

^{195}Pt. Also for Pt many constraints had to be made in order to obtain reasonable fits of the observed A values to the effective a parameters. $a^{01}(5d^9 6s)$ and $a^{12}(5d^9 6s)$ were related to the parameters of the configuration $5d^8 6s^2$ using the ratio of the spin-orbit constants, and the 5d-electron contact parameters a^{10}_{5d} were assumed to be equal for the two configurations. The remaining four a's were fitted to eleven experimental A factors with a mean deviation of 12% (neglecting the $^3P_1$ state for which the A factor has a very large experimental uncertainty).

^{197}Au. From the A factor of the $^2S_{1/2}$ ground state one immediately evaluates a^{10}_{6s}. For the $5d^9 6s^2$ ^2D multiplet $<r^{-3}>^{12}$ has been estimated from the results for the neighboring elements by linear interpolation. Using this value a^{01} and a^{10}_{5d} could be determined from the experimental A factors. The ratio b^{13}/b^{11} was fixed at the MCDF value; then, b^{02} and b^{13} were evaluated from the B factors of the ^2D multiplet.

This detailed analysis shows that the effective operator formalism works quite well, even in the case of strong deviations from pure SL coupling, although many approximations had to be made. On the other hand, the experimental hfs constants cannot be reproduced within their relative experimental error limits of 10^{-6}-10^{-4}. Typical deviations of the recalculated hfs constants from the experimental values are some parts in 10^{-2}; for the heavier elements of the 5d-shell the deviations reach the order of even 10^{-1}. These deviations are partly due to deficiencies of the intermediate coupling wavefunctions, and partly the result of neglecting the hfs interaction between the configurations of the model space $(d+s)^N$ and of neglecting SL-dependent CI effects.

Therefore, in order to resolve the discrepancies, the following improvements are very desirable in future investigations:

1) The intermediate coupling wavefunctions have to be improved by taking into account effective three-particle fine-structure interactions.

2) Effective radial parameters of the hyperfine interaction between the configurations of the $(d+s)^N$ model space have to be included in the fitting procedure.

3) SL-dependent CI effects must be taken into account by introducing additional effective radial parameters as was done for the 3d-shell atoms /31,32/. For these purposes, however, the hfs has to be measured for many states of the $(d+s)^N$ configurations of an atom. This will probably become feasable by use of single mode cw dye lasers (Sect.3.3).

6.2 Comparison Between Experimental and Theoretical Radial Integrals

According to (9a,b) the radial expectation values $<r^{-3}>^{k_s k_1}$ may be evaluated from the $a^{k_s k_1}$ parameters,

$$<r^{-3}>^{k_s k_1} = 0.010481 \; \frac{I}{\mu_I} \; a^{k_s k_1} \tag{76}$$

where $a^{k_s k_1}$ is given in MHz, μ_I in nuclear magnetons, and $<r^{-3}>^{k_s k_1}$ in units of a_0^{-3}. In Tables 9 and 10 these experimental $<r^{-3}>$ values are summarized and compared with relativistic MCDF values. Such $<r^{-3}>$ values which were fixed or whose ratio to another parameter was fixed are not given in the tables. Because of the lack of independently measured values of the electric quadrupole moments only the relative size of the corresponding $<r^{-3}>$ values can be compared with relativistic results.

According to (17,18), relativistic $<r^{-3}>$ values have been evaluated for the configurations $4d^{N-1}5s$, $4d^{N-2}5s^2$, $5d^{N-1}6s$, and $5d^{N-2}6s^2$ from relativistic self-consistent-field radial wavefunctions F_{nlj} and G_{nlj} (Tables 9 and 10). These radial wavefunctions have been calculated by using the atomic MCDF program developed by GRANT et al. /43/. For averaging over all the jj subconfigurations arising from the entire SL configurations $4d^{N-1}5s$, $4d^{N-2}5s^2$, $5d^{N-1}6s$, and $5d^{N-2}6s^2$, respectively, the extended average level (EAL) method has been applied. The exchange potentials were included in the self-consistent procedure without any approximations. The nucleus has been treated as a point charge distribution in the case of the 4d-shell atoms and as a uniformly charged sphere of radius $R_{nuc} = 1.2 \cdot A^{1/3}$ fm for the 5d-shell atoms

Table 9. Experimental and theoretical $\langle r^{-3}\rangle$ values (in units of a_0^{-3}) for 4d-shell atoms

Atom and configuration	Method	Magnetic dipole interaction				Electric quadrupole interaction			
		$\langle r^{-3}\rangle_{01}$	$\langle r^{-3}\rangle_{12}$	$\langle r^{-3}\rangle_{4d}$	$\langle r^{-3}\rangle_{5s}$	$\langle r^{-3}\rangle_{02}$	$\dfrac{\langle r^{-3}\rangle_{13}}{\langle r^{-3}\rangle_{02}}$	$\dfrac{\langle r^{-3}\rangle_{11}}{\langle r^{-3}\rangle_{02}}$	$\dfrac{\langle r^{-3}\rangle_{11}}{\langle r^{-3}\rangle_{13}}$
Y $4d5s^2$	MCDF	1.62	1.70	-0.037		1.63	0.112	-0.040	-0.358
	Exp		0.27	-0.770					
$4d^2s$	MCDF	1.29	1.37	-0.034	26.9	1.30	0.121	-0.045	-0.378
Zr $4d^2 5s^2$	MCDF	2.31	2.44	-0.057		2.32	0.120	-0.043	-0.356
	Exp		1.45	-0.854			0.124	-0.028	-0.223
$4d^3s$	MCDF	1.96	2.08	-0.055	31.3	1.97	0.129	-0.048	-0.375
	Exp	1.93	1.09	-1.375	38.5		0.036	-0.157	-4.313
Nb $4d^3 5s^2$	MCDF	3.04	3.21	-0.079		3.05	0.128	-0.046	-0.355
	Exp	3.04	1.83	-1.734			0.116	-0.028	-0.239
$4d^4 5s$	MCDF	2.66	2.83	-0.079	35.5	2.67	0.138	-0.051	-0.372
	Exp	2.70	2.12	-3.927	51.4		0.119	-0.011	-0.095
Mo $4d^4 5s^2$	MCDF	3.82	4.05	-0.105		3.84	0.137	-0.048	-0.352
	Exp		2.30	-2.016			0.241	-0.034	-0.143
$4d^5 5s$	MCDF	3.41	3.65	-0.107	39.6	3.43	0.146	-0.054	-0.369
	Exp			-3.215	52.1				
Tc $4d^5 5s^2$	MCDF	4.66	4.96	-0.136		4.69	0.146	-0.051	-0.351
	Exp			-0.910					
$4d^6 5s$	MCDF	4.22	4.53	-0.139	43.7	4.25	0.155	-0.057	-0.366
	Exp	4.02	4.02	-2.624	54.9				
Ru $4d^6 5s^2$	MCDF	5.57	5.95	-0.172		5.61	0.155	-0.054	-0.349

$4d^7 5s$	MCDF	5.10	5.49	-0.177	47.8	5.13	0.164	-0.060	-0.365
	Exp	4.92	3.81	-2.329	60.9		0.098	-0.132	-1.342
Rh $4d^7 5s^2$	MCDF	6.55	7.03	-0.213		6.60	0.164	-0.057	-0.348
$4d^8 5s$	MCDF	6.04	6.54	-0.220	51.9	6.09	0.173	-0.063	-0.363
	Exp	5.96	3.98	-0.763	60.0				
Pd $4d^8 5s^2$	MCDF	7.60	8.19	-0.260		7.67	0.173	-0.060	-0.347
$4d^9 5s$	MCDF	7.07	7.67	-0.270	56.0	7.12	0.183	-0.066	-0.360
	Exp	7.40		-2.665	73.6				
Ag $4d^9 5s^2$	MCDF	8.74	9.45	-0.315		8.82	0.183	-0.063	-0.346
	Exp	9.01		-1.602					
$4d^{10} 5s$	MCDF				60.2				
	Exp				79.8				

Table 10. Experimental and theoretical $\langle r^{-3}\rangle$ values (in units of a_0^{-3}) for 5d-shell atoms

Atom and configuration	Method	Magnetic dipole interaction				Electric quadrupole interaction			
		$\langle r^{-3}\rangle_{01}$	$\langle r^{-3}\rangle_{12}$	$\langle r^{-3}\rangle_{5d}^{10}$	$\langle r^{-3}\rangle_{6s}^{10}$	$\langle r^{-3}\rangle_{02}$	$\dfrac{\langle r^{-3}\rangle_{13}}{\langle r^{-3}\rangle_{02}}$	$\dfrac{\langle r^{-3}\rangle_{11}}{\langle r^{-3}\rangle_{02}}$	$\dfrac{\langle r^{-3}\rangle_{11}}{\langle r^{-3}\rangle_{13}}$
Lu 5d6s²	MCDF	3.03	3.70	-0.296		3.11	0.523	-0.173	-0.331
	Exp		-1.30	-0.632			0.360	-0.123	
5d²6s	MCDF	2.47	3.08	-0.272	96.5	2.54	0.556	-0.191	-0.344
	Exp	2.62		-7.461	135.5				
Hf 5d²6s²	MCDF	4.23	5.20	-0.423		4.36	0.539	-0.177	-0.328
	Exp	4.09	0.46	-0.134			0.499	-0.189	-0.380
5d³6s	MCDF	3.66	4.59	-0.412	113.6	3.76	0.574	-0.196	-0.342
Ta 5d³6s²	MCDF	5.41	6.67	-0.552		5.58	0.556	-0.181	-0.326
	Exp	5.09	1.47	-0.879			0.458	-0.044	-0.095
5d⁴6s	MCDF	4.82	6.06	-0.550	131.2	4.96	0.589	-0.199	-0.338
	Exp	5.91		13.85	82.2				
W 5d⁴6s²	MCDF	6.62	8.20	-0.691		6.83	0.573	-0.186	-0.324
	Exp		3.38	-2.751					
5d⁵6s	MCDF	6.00	7.58	-0.694	149.2	6.19	0.604	-0.203	-0.335
	Exp				149.5				
Re 5d⁵6s²	MCDF	7.87	9.80	-0.839		8.13	0.590	-0.190	-0.322
	Exp	6.90	7.90	-1.237			0.586	-0.065	-0.111
5d⁶6s	MCDF	7.23	9.16	-0.851	168.0	7.46	0.621	-0.207	-0.333
	Exp	6.41		1.526	-151.2				
Os 5d⁶6s²	MCDF	9.18	11.48	-1.000		9.50	0.608	-0.195	-0.320

Element	Configuration	Method								
	$5d^7 6s$	Exp		3.34						
		MCDF	8.52	10.84	-3.013		8.80	0.517	-0.242	-0.468
Ir	$5d^7 6s^2$	Exp	10.56	13.27	-1.019	187.6		0.638	-0.211	-0.330
		MCDF	13.78	18.13	-1.176	186.7	10.94	0.626	-0.199	-0.319
	$5d^8 6s$	Exp	9.87	12.61	-10.207	208.1		0.404	-0.096	-0.239
		MCDF	12.01	15.16	-1.201	255.4	10.22	0.654	-0.215	-0.328
Pt	$5d^8 6s^2$	MCDF		4.50	-1.367		12.46	0.644	-0.204	
	$5d^9 6s$	Exp	11.62	14.48	-2.293	229.7		0.672	-0.219	-0.317
		MCDF	11.29	17.16	-1.398	256.7	11.70	0.663	-0.209	-0.326
Au	$5d^9 6s^2$	MCDF	13.53		-1.574		14.06	0.532	-0.176	-0.315
		Exp	13.92		-5.982					
	$5d^{10} 6s$	MCDF				252.5				
		Exp				334.2				

for which the effect of the finite nuclear charge distribution on the 6s-electron contact interaction is of the order of some parts in 10^{-2} and cannot be neglected.

First, some characteristics of the relativistic results shall be discussed. The results for the configurations $d^{N-1}s$ show the same general trend as those obtained for the $d^{N-2}s^2$ configurations. For a given N the values of $\langle r^{-3}\rangle^{01}$, $\langle r^{-3}\rangle^{12}$, and $\langle r^{-3}\rangle^{02}$ are, however, smaller in the configuration $d^{N-1}s$ than in the configuration $d^{N-2}s^2$, which means that the nd-orbitals are more contracted when two s-electrons are present. Within each series the d- as well as the s-electrons are bound stronger with increasing N, which gives an increase in the $\langle r^{-3}\rangle^{01}$, $\langle r^{-3}\rangle^{12}$, $\langle r^{-3}\rangle^{02}$, and $\langle r^{-3}\rangle^{10}_s$ values. For 5d-shell atoms the ratios $\langle r^{-3}\rangle^{13}/\langle r^{-3}\rangle^{02}$ and $\langle r^{-3}\rangle^{11}/\langle r^{-3}\rangle^{02}$ are larger by a factor of about four than for 4d-shell atoms, indicating the importance of relativistic effects in 5d-shell atoms. These ratios increase slightly with increasing N, while the ratio $\langle r^{-3}\rangle^{11}/\langle r^{-3}\rangle^{13}$ is nearly constant over the whole shell and differs by only 4% between the configurations $d^{N-1}s$ and $d^{N-2}s^2$.

Comparing these relativistic results with the experimental $\langle r^{-3}\rangle$ values, the influence of CI effects on the hyperfine interaction can be estimated. Figures 9-11 describe graphically the variation of the effective radial integrals of the magnetic dipole interaction as functions of the atomic number for the series $4d^{N-1}5s$, $4d^{N-2}5s^2$, $5d^{N-1}6s$, and $5d^{N-2}6s^2$. Since most of the d-electron $\langle r^{-3}\rangle$ values for the configurations $5d^{N-1}6s$ were related to the respective parameters of the configurations $5d^{N-2}6s^2$, there are only very few independent experimental data for this series, and consequently only the graph of the 6s-electron contact parameter is given.

Several conclusions can be drawn from Tables 9 and 10 and Figs.9-11.

1) In all configurations for which all three $\langle r^{-3}\rangle$ values of the electric quadrupole interaction could be determined from the experimental data the signs of the ratios $\langle r^{-3}\rangle^{13}/\langle r^{-3}\rangle^{02}$, $\langle r^{-3}\rangle^{11}/\langle r^{-3}\rangle^{02}$, and $\langle r^{-3}\rangle^{11}/\langle r^{-3}\rangle^{13}$ are consistent with the relativistic predictions. The absolute values of these ratios show, however, some irregular scattering. Except for the $4d^3 5s$ configuration of Zr, the value of $\langle r^{-3}\rangle^{13}/\langle r^{-3}\rangle^{02}$ differ by less than a factor of two from the Dirac-Fock values, while the ratios $\langle r^{-3}\rangle^{11}/\langle r^{-3}\rangle^{02}$ and $\langle r^{-3}\rangle^{11}/\langle r^{-3}\rangle^{13}$ show deviations of up to a factor of four. Since the effective radial parameters b^{13} and b^{11} are very sensitive to the intermediate coupling wavefunctions, these differences between relativistic and experimental values for the ratios of the electric quadrupole interaction radial integrals are presumably due to slight deficiencies of the eigenvectors. For this reason values of the electric quadrupole moments Q are usually evaluated from the parameter b^{02} (Sect.7.2).

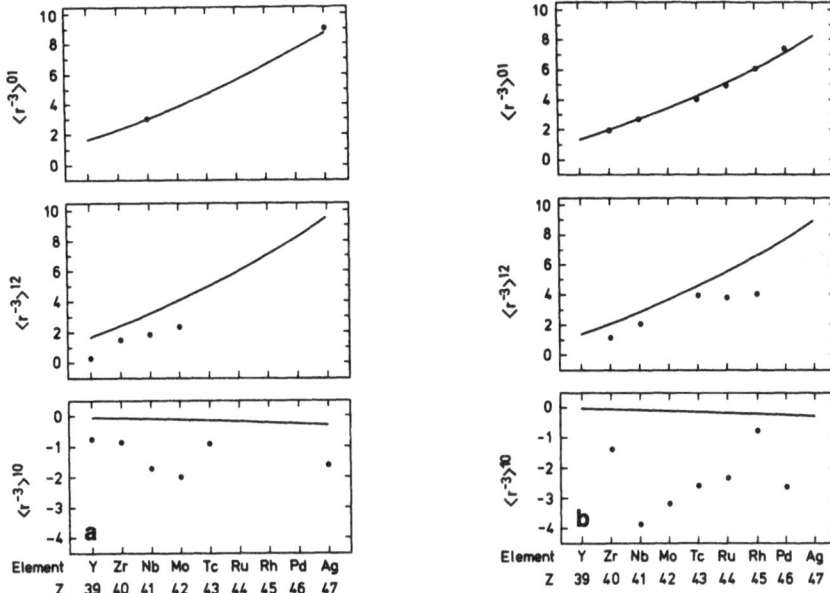

Fig.9a,b. Comparison between experiment and theory for the radial parameters of the (a) $4d^{N-2}5s^2$ and (b) $4d^{N-1}5s$ series ($<r^{-3}>$ in units of a_0^{-3})

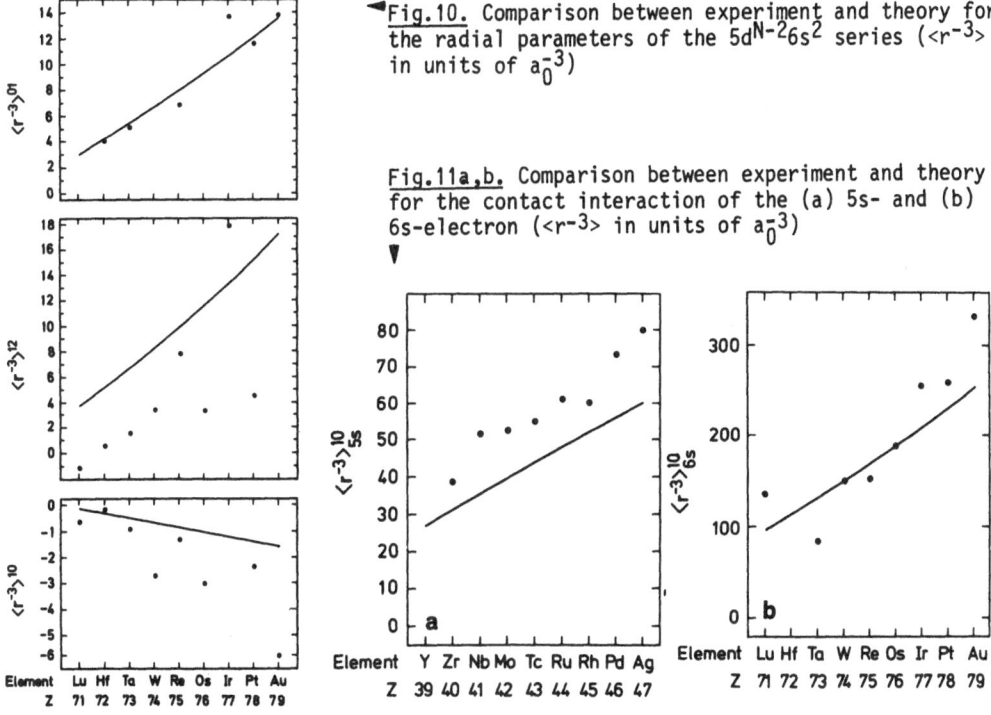

◄**Fig.10.** Comparison between experiment and theory for the radial parameters of the $5d^{N-2}6s^2$ series ($<r^{-3}>$ in units of a_0^{-3})

Fig.11a,b. Comparison between experiment and theory for the contact interaction of the (a) 5s- and (b) 6s-electron ($<r^{-3}>$ in units of a_0^{-3})

2) The experimental values for $\langle r^{-3}\rangle^{01}$ are in rather good agreement with the relativistic values. The deviations are of the order of a few percent. This result justifies the approximation $a^{01} = a^{01}_{MCDF}$ made for some configurations when fitting the experimental magnetic dipole hfs constants to the a parameters. On the other hand, the systematic deviation of the experimental $\langle r^{-3}\rangle^{12}$ values from the MCDF results is appreciable although the parameter a^{12} is very sensitive to the intermediate coupling eigenvectors, too. This clearly indicates that $\langle r^{-3}\rangle^{12}$ is strongly influenced by CI, while $\langle r^{-3}\rangle^{01}$ is fairly insensitive. The effects of CI generally lessen the $\langle r^{-3}\rangle^{12}$ values, and this effect is more pronounced for 5d-shell atoms than for 4d-shell atoms. The same influence, but still smaller, has been found for atoms with an open 3d-shell /58/.

3) The systematic disagreement of an order of magnitude for the 4d-electron contact parameter $\langle r^{-3}\rangle^{10}_{4d}$ shows that the contact part of the interaction cannot be explained by relativistic effects and must be caused by spin polarization. For the 5d-shell atoms the relativistic effective radial integrals $\langle r^{-3}\rangle^{10}_{5d}$ vary from $-0.3\ a_0^{-3}$ to $-1.6\ a_0^{-3}$. For the $5d^{N-2}6s^2$ configurations the experimental $\langle r^{-3}\rangle^{10}_{5d}$ parameters are smaller than the relativistic results indicating the influence of spin polarization, although the experimental values scatter from $-0.1\ a_0^{-3}$ to $-10\ a_0^{-3}$. This scattering is possibly due to mathematical correlations in the least-squares fit procedure between the contact parameters of the 5d- and the 6s-electrons, a^{10}_{5d} and a^{10}_{6s}, respectively. Such correlations are even more important for the configurations $5d^{N-1}6s$. The magnetic dipole interaction of the unpaired 6s-electron clearly dominates the 5d contact interaction leading to quite unrealistic $a^{10}_{5d}(5d^{N-1}6s)$ values. Further levels belonging to $5d^{N-1}6s$ configurations have to be investigated in order to obtain physically reliable results for the d-electron contact interaction in these configurations. Unfortunately, there are no systematic calculations of spin polarization effects for 4d- and 5d-shell atoms. For 3d-shell atoms BAGUS et al. /205/ could reproduce the general trend of the experimental $\langle r^{-3}\rangle^{10}_{3d}$ values /206/ using the SUHF method.

4) The experimental values of $\langle r^{-3}\rangle^{10}_{5s}$ are systematically larger than the relativistic values indicating the influence of CI on the s-electron contact interaction. The same situation has been established already for the contact parameter $\langle r^{-3}\rangle^{10}_{4s}$ in 3d-shell atoms /58/. In the case of the $5d^{N-1}6s$ configurations the slope of the experimental graph seems to be different from that of the calculated one, although the experimental trend has to be clarified. For this purpose it is necessary to measure the contact interaction also for the $5d^36s$ configuration in Hf and to check

the $<r^{-3}>_{6s}^{10}$ value for Lu which disagrees with the general trend of the 6s contact interaction parameters. Furthermore, the Bohr-Weisskopf correction ε_{BW} (Sect.2.8), which is expected to be of the order of some parts in 10^{-2} for the 5d-shell atoms, affects the $<r^{-3}>_{ns}^{10}$ values. Because of the lack of reliable calculations of ε_{BW} the results have not been corrected for this effect. Therefore, in addition to further measurements detailed calculations of the influence of the magnetization distribution in the nucleus on the contact interaction have to be carried out.

7. Nuclear Moments and Hyperfine Anomalies

7.1 Direct Measurement of Magnetic Dipole Moments

In order to determine the nuclear magnetic dipole moments of the nuclei with the
ABMR method independently of the interaction between the nucleus and the atomic
electrons, one has to extract from adequate hfs measurements the direct interaction between the nuclear magnetic dipole moment and the external magnetic field
applied in the C-field region of the ABMR apparatus.

According to the Hamiltonian of (54) the resonance frequency of an rf transition
is a function of the magnetic field H, the nuclear and electronic g factors g_I and
g_J, and the hfs constants A_k (k = 1,..., min(2I,2J)) of the state in which the transition is observed,

$$\nu = \nu(H, g_I, g_J, A_k) \quad . \tag{77}$$

Expanding ν about approximated values H^0, g_I^0, g_J^0, and A_k^0, and assuming that second-
and higher-order terms can be ignored one finds

$$\nu = \nu(H^0, g_I^0, g_J^0, A_k^0) + \frac{\partial \nu}{\partial H}\bigg|_0 dH + \frac{\partial \nu}{\partial g_I}\bigg|_0 dg_I + \frac{\partial \nu}{\partial g_J}\bigg|_0 dg_J + \sum_k \frac{\partial \nu}{\partial A_k}\bigg|_0 dA_k \tag{78}$$

where $dH = H-H^0$, $dg_I = g_I-g_I^0$, $dg_J = g_J-g_J^0$, and $dA_k = A_k-A_k^0$. The symbol $\big|_0$ indicates
that the derivatives have to be taken at the point $H = H^0$, $g_I = g_I^0$, $g_J = g_J^0$, $A_k = A_k^0$.
It is possible to evaluate $\nu(H^0, g_I^0, g_J^0, A_k^0)$ by diagonalizing the matrix of the Hamiltonian $H_{hfs,Z}$ (Sect.2.6).

Considering now two transitions ν_1 and ν_2 belonging to the same atomic state and
measured at the same magnetic field H, one constructs the local field-independent
quantity

$$d\nu = \left[\nu_1-\nu_1(H^0, g_I^0, g_J^0, A_k^0)\right] - \frac{\alpha_1}{\alpha_2}\left[\nu_2-\nu_2(H^0, g_I^0, g_J^0, A_k^0)\right] \tag{79}$$

where $\alpha = \frac{\partial \nu}{\partial H}\big|_0$ and $|\alpha_1/\alpha_2| \leq 1$. Combining (78) and (79) one finds

$$dv = \left(\frac{\partial v_1}{\partial g_I} \bigg|_0 - \frac{\alpha_1}{\alpha_2} \frac{\partial v_2}{\partial g_I} \bigg|_0 \right) dg_I + (\dots) dg_J + \sum_k (\dots) dA_k \quad . \tag{80}$$

If the electronic g_J factor and the hfs interaction constants A_k have been measured previously with such accuracy that the terms comprising dg_J and dA_k can be neglected, the nuclear g_I factor can be derived from the expression

$$g_I = g_I^0 + \frac{dv}{(\mu_B/h) \, H^0 (1-g_I^0/g_J^0)^{-1} \left(\sigma_1 - \frac{\alpha_1}{\alpha_2} \sigma_2 \right)} \quad , \tag{81}$$

where we have used the relation /207/

$$\frac{\partial v}{\partial g_I} = \frac{(\mu_B/h) g_J \, \sigma-\alpha}{g_J-g_I} \tag{82}$$

and σ is the polarization of the transition.

This consideration allows us to decide whether a frequency pair (v_1,v_2,H) is suited to determine the g_I factor within a given uncertainty. The influence of the uncertainties in g_J and in the hfs constants A_k on the g_I error Δg_I as well as the contributions of the experimental uncertainties in the resonance frequencies v_1 and v_2 to Δg_I must be smaller than the desired g_I error. Furthermore, according to the available rf power the transition matrix element $<\mathscr{F}M_F \mid g_J J_\pm + g_I I_\pm \mid \mathscr{F}' M_F'>$ has to be sufficiently large. Since both transitions should be induced at the same magnetic field, that means, in the same rf loop, we have the additional condition $|\sigma_1| = |\sigma_2|$ = 1. Those transition pairs (v_1,v_2,H) which are suited to determine the g_I factor with a given uncertainty are selected by a computer program /208/, which evaluates for all possible frequency pairs at several magnetic fields the different contributions to Δg_I.

In general, transitions of the type $\Delta M_J = 0$, $\Delta M_I = \pm 1$ measured at high magnetic fields ($H \gtrsim 1000$ Oe) turn out to be best suited. Such transitions are, however, not detectable directly but require application of the triple-resonance technique /209, 210/, i.e., two auxiliary rf transitions $v_a(A)$ and $v_a(B)$ of the type $\Delta M_J \neq 0$ are induced in two homogeneous magnetic fields C_A and C_B which are separated from the original C field.

Up to now the nuclear magnetic dipole moments of the following isotopes of refractory elements have been measured by this method: ^{99,101}Ru, ^{177,179}Hf, and ^{191,193}Ir. In Fig.12 the two transition pairs selected for the $^5F_5$ ground state of ^{99,101}Ru and the appropriate auxiliary transitions are shown in the hfs level diagram. Figure 13 shows a block diagram of the rf setup for these measurements. Because $H(C_A) = H(C_B) \cong 20$ Oe, the auxiliary frequencies $v_a(A) = v_a(B)$ were of the

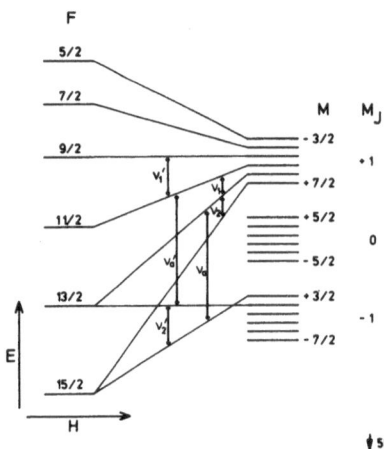

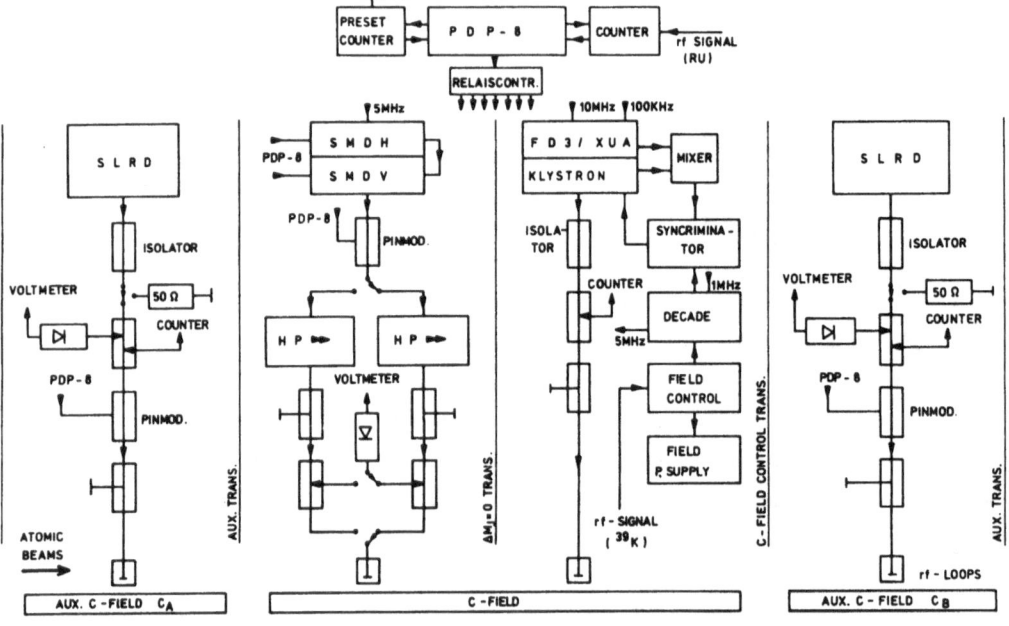

Fig.12. Schematic hfs level diagram for the ground state $^5F_5$ of ^{99,101}Ru with the measured transition pairs ν_1, ν_2 and ν_1', ν_2'. ν_a and ν_a' are the respective auxiliary transitions

Fig.13. Block diagram of the rf setup for the g_I measurements

order of the hfs splitting. One of the two rf loops inside the C-field served for the field control using a strongly field-dependent transition in ^{39}K (Sect.3.1), while the other one was used to induce the two $\Delta M_J = 0$ transitions ν_1 and ν_2 by turns.

A serious problem arises from the fact that hfs and Zeeman interactions with other atomic levels (Sect.2.7) become extremely important at a high magnetic field. The "field-field" term of the off-diagonal corrections is usually very small and may be neglected. While the "hfs-hfs" term can be taken into account by using the effective hfs interaction constants measured at low magnetic fields, the "hfs-field" term leads to a "pseudo-moment" /211/ and has to be calculated using second-order perturbation theory. The influence of the off-diagonal interactions on g_I was up to 10% for ^{99,101}Ru and ^{191,193}Ir and about 1% for ^{177,179}Hf.

In order to have an experimental test of the calculated corrections, several transition pairs with corrections as different as possible have been selected for the g_I measurement. For ^{99,101}Ru and ^{191,193}Ir measurements were performed in several atomic states. Because of the excellent agreement between all measurements after applying the off-diagonal corrections it was possible to determine the nuclear magnetic dipole moments of ^{177,179}Hf with an accuracy of better than 0.2% and those of ^{99,101}Ru with an accuracy of about 1% /160,212/,

$$\mu_I(^{177}\text{Hf}) = 0.7836\ (6)\mu_N, \qquad \mu_I(^{179}\text{Hf}) = -0.6329(13)\mu_N,$$
$$\mu_I(^{99}\text{Ru}) = -0.6381(51)\mu_N, \qquad \mu_I(^{101}\text{Ru}) = -0.7152(60)\mu_N.$$

In the case of ^{191,193}Ir five transition pairs have been measured in the atomic ground state $^4F_{9/2}$ at magnetic fields up to 2100 Oe. The results agree within the limits of error. As weighted averages one finds /181/

$$\mu_I(^{191}\text{Ir}) = 0.1485(6)\mu_N, \qquad \mu_I(^{193}\text{Ir}) = 0.1613(6)\mu_N.$$

The off-diagonal "hfs-field" corrections turned out to vary considerably when different sets of effective radial parameters $a^{k}s^{k}1$ and $b^{k}s^{k}1$ were used in the second-order calculations. Therefore, an uncertainty of 50% has been assumed for the corrections. Measurements in the atomic states $J = 3/2$ (4079 cm^{-1}) and $J = 5/2$ (5785 cm^{-1}) led to values for the magnetic moments which are about 10% smaller than those obtained from the atomic ground state. Since the off-diagonal effects are larger than for the ground state and since these states are very far from SL coupling, the results have not been taken into account. This discrepancy, however, clearly indicates the present limitation of this method for the measurement of nuclear magnetic dipole moments in complex atoms.

7.2 Electric Quadrupole Moments

According to (9c) the nuclear electric quadrupole moments may be evaluated from the b^{02} parameters,

$$Q = 0.00425595\ \frac{b^{02}}{\langle r^{-3}\rangle^{02}} \tag{83}$$

where b^{02} is given in MHz, $\langle r^{-3}\rangle^{02}$ in units of a_0^{-3}, and Q in barns. If one uses the relativistic value for $\langle r^{-3}\rangle^{02}$, one finds from (83) the uncorrected quadrupole moment

$$Q'_{hfs} = Q(1+\Delta^{02}) \quad .$$

The corrected quadrupole moment Q can only be determined if the CI effect Δ^{02} on $\langle r^{-3}\rangle^{02}$ can be estimated (Sect.2.4). Another procedure in evaluating the quadrupole moments is to use a value for the ratio $\langle r^{-3}\rangle^{01}/\langle r^{-3}\rangle^{02}$ and to express $\langle r^{-3}\rangle^{01}$ according to (76),

$$Q = 0.406063 \, \frac{\mu_I}{I} \, \frac{b^{02}}{a^{01}} \, \frac{\langle r^{-3}\rangle^{01}}{\langle r^{-3}\rangle^{02}} \quad . \tag{84}$$

Using a relativistic value for the ratio $\langle r^{-3}\rangle^{01}/\langle r^{-3}\rangle^{02}$ leads to an uncorrected value for the quadrupole moment, too,

$$Q_{hfs} = Q(1+\Delta^{02}-\Delta^{01}) \quad .$$

In this case one has to know the difference $(\Delta^{02}-\Delta^{01})$ of the CI effects on $\langle r^{-3}\rangle^{02}$ and $\langle r^{-3}\rangle^{01}$ for the determination of Q.

In order to obtain reliable values for the quadrupole moments the a^{01} and b^{02} parameters should be determined in multi-parameter fits from as many A and B factors as possible, because hfs fits using the nonrelativistic approximation and neglecting CI effects lead in many cases, especially for 5d-shell atoms, to physically unreasonable parameter values. This condition is fulfilled for most of the 4d- and 5d-shell atoms due to the extensive hfs measurements in these atoms. However, since there are no theoretical values of Δ^{02} and Δ^{01} available, only uncorrected quadrupole moments can be determined from the experiments. Tables 11 and 12 summarize the values Q_{hfs} evaluated from the experimental ratios b^{02}/a^{01} (Tables 7,8) and from the relativistic ratios $\langle r^{-3}\rangle^{01}/\langle r^{-3}\rangle^{02}$ (Tables 9,10). The uncertainty of these uncorrected quadrupole moments is mainly due to the uncertainty in the ratio b^{02}/a^{01}. Because of the different qualities of the parameter fits an error of 5% has been assumed for 4d-isotopes, while 10% and 20% uncertainties have been adopted for the $5d^{N-2}6s^2$ and the $5d^{N-1}6s$ series, respectively.

The nuclides of the 5d-shell atoms lie between the strongly deformed rare-earth region and the spherical lead region. This is reflected by the quadrupole moments which vary from 0.6 b for ^{197}Au to about 6 b for ^{175}Lu. In Table 12 the quadrupole moments determined from (84) (Q_{hfs}) are compared with results obtained from the hfs of mesonic atoms (Q_{mes}) and with quadrupole moments (Q') calculated from the in-

Table 11. Quadrupole moments for some 4d isotopes determined from the hyperfine structure

Isotope	I	Q_{hfs} [b]
^{91}Zr	5/2	-0.206 (10)[a]
^{93}Nb	9/2	-0.366 (18)[a]
^{95}Mo	5/2	-0.022 (1)[b]
^{97}Mo	5/2	0.255 (13)[a]
^{99}Tc	9/2	-0.129 (6)[a]
^{99}Ru	5/2	0.079 (4)[c]
^{101}Ru	5/2	0.457 (23)
^{105}Pd	5/2	0.647 (32)

[a] Average of the results obtained for the configurations $4d^{N-2}5s^2$ and $4d^{N-1}5s$.

[b] From $Q(^{97}Mo)$ and $Q(^{95}Mo)/Q(^{97}Mo)$ = -11.5 /144/.

[c] From $Q(^{101}Ru)$ and $Q(^{99}Ru)/Q(^{101}Ru)$ = 0.172 /160/.

trinsic quadrupole moments Q_0 using the well-known projection factor /18/ between the intrinsic and the spectroscopic quadrupole moments. From the ratios Q_{hfs}/Q_{mes} or Q_{hfs}/Q' the difference of the CI effects on $<r^{-3}>^{02}$ and $<r^{-3}>^{01}$ can be estimated (last column of Table 12). Figure 14 shows this difference, which represents an "experimental" shielding (Sternheimer) factor, for the $5d^{N-2}6s^2$ series. Despite the fact that the results suffer from severe deficiencies in the assumptions being used throughout the analysis of the experimental data, there is a decreasing trend of the CI effect $(\Delta^{02}-\Delta^{01})$ from Lu to Au. Therefore it seems very desirable and rewarding to perform explicit calculations of such effects using many-body techniques. Enough experimental data have been obtained to justify renewed theoretical efforts. Furthermore, measurements of the hfs in still more atomic states together with a more elaborate analysis of the data taking into account SL-dependent CI effects may lead to a direct experimental determination of the shielding parameters Δ^{02} and Δ^{01}.

On the other hand, because the evaluation of the quadrupole moments from the hfs of mesonic atoms is nearly model independent, the values of Q_{hfs}/Q_{mes} may be used to estimate the shielding effects for such atoms for which no mesonic hfs has been measured so far (Hf, Os). By linear interpolation one finds $(\Delta^{02}-\Delta^{01})_{Hf}$ = 0.30(12) and $(\Delta^{02}-\Delta^{01})_{Os}$ = 0.07(16). Thus, the following quadrupole moments corrected for shielding effects can be given: $Q(^{177}Hf)$ = 3.30(65) b, $Q(^{179}Hf)$ = 3.72(74) b, and $Q(^{189}Os)$ = 0.88(18) b.

Table 12. Comparison of quadrupole moments for some 5d isotopes determined from the hfs of neutral atoms, from the hfs of mesonic atoms, and by nuclear methods

Isotope	I	Config.	Q_{hfs} [b]	Q_{mes} [b][a]	Q' (from Q_0) [b][b]	$\Delta^{02} - \Delta^{01}$ [c]
^{175}Lu	7/2	$5d6s^2$	4.83 (48)	3.46 (6)	3.75 (14)	0.40 (14)
		$5d^26s$	6.3 (1.3)			0.83 (37)
^{177}Hf	7/2	$5d^26s^2$	4.71 (47)[d]		3.55 (12)	0.33 (14)
^{179}Hf	9/2	$5d^26s^2$	5.32 (53)		3.82 (13)	0.39 (15)
^{181}Ta	7/2	$5d^36s^2$	3.93 (39)	3.28 (6)	3.15 (6)	0.20 (12)
		$5d^46s$	3.43 (69)			0.05 (21)
^{185}Re	5/2	$5d^56s^2$	2.90 (29)[e]	2.19 (2)	2.14 (6)	0.32 (13)
		$5d^66s$	2.72 (54)[e]			0.24 (25)
^{187}Re	5/2	$5d^56s^2$	2.74 (27)	2.08 (2)	2.05 (6)	0.32 (13)
		$5d^66s$	2.57 (51)			0.24 (25)
^{189}Os	3/2	$5d^66s^2$	0.95 (10)		0.82 (3)	0.16 (13)
		$5d^76s$	0.75 (15)			-0.09 (19)
^{191}Ir	3/2	$5d^76s^2$	0.70 (7)[f]	0.86 (4)	0.87 (5)	-0.19 (9)
^{193}Ir	3/2	$5d^76s^2$	0.63 (6)	0.78 (4)	0.80 (3)	-0.19 (9)
^{197}Au	3/2	$5d^96s^2$	0.58 (6)	0.547 (16)		0.06 (11)

[a] The values for the quadrupole moments determined from the hfs of mesonic atoms have been taken from /213/ for ^{175}Lu, /214/ for ^{181}Ta and ^{185,187}Re, /215/ for ^{191,193}Ir, and /216/ for ^{197}Au.

[b] The values for Q_0 have been taken from /217/.

[c] Calculated from Q_{hfs} and Q_{mes}, and from Q_{hfs} and Q' if no value from mesonic atoms is available.

[d] From Q(^{179}Hf) and Q(^{177}Hf)/Q(^{179}Hf) = 0.885 /169/.

[e] From Q(^{187}Re) and Q(^{185}Re)/Q(^{187}Re) = 1.0567 /177/.

[f] From Q(^{193}Ir) and Q(^{191}Ir)/Q(^{193}Ir) = 1.1055 /181/.

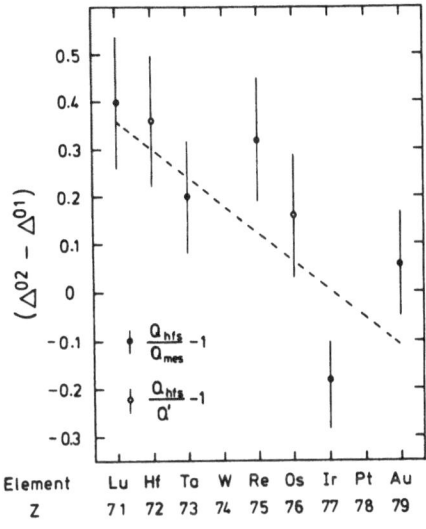

Fig.14. Difference of CI effects on $<r^{-3}>^{02}$ and $<r^{-3}>^{01}$ for the $5d^{N-2}6s^2$ series. The dashed line has been obtained by a linear least-squares fit of the values derived from Q_{hfs} and Q_{mes} (•)

Figure legend within plot:
$\frac{Q_{hfs}}{Q_{mes}} - 1$

$\frac{Q_{hfs}}{Q'} - 1$

Element	Lu	Hf	Ta	W	Re	Os	Ir	Pt	Au
Z	71	72	73	74	75	76	77	78	79

7.3 Hyperfine Anomalies

The hyperfine anomaly has been determined for five pairs of odd-A isotopes of 4d and 5d elements. Using (66) the s-electron hyperfine anomalies $^1\Delta_s^2$ were evaluated from the experimental values $^1\Delta_{exp}^2$. For this purpose the contact parts A_s of the A factors have been calculated from the expressions of the A factors as linear combinations of the $a^{k_s k_l}$ parameters (Sect.2.5) and from the best-fit values of the $a^{k_s k_l}$ parameters (Tables 7,8). The results are summarized in Table 13. In the cases of Re and Ir the quoted values are the averages from measurements in six metastable states.

For Ir and Au application of the Moskowitz-Lombardi rule /116/ yields hyperfine anomalies which agree very well with the experimental ones. In the case of Ag FUJITA and ARIMA /117/ calculated the hyperfine anomaly by taking into account effects of nuclear core polarization and mesonic exchange currents. Their result is in good agreement with the experimental value.

For ^{95,97}Mo and ^{185,187}Re the hyperfine anomalies are very small and the Breit-Rosenthal effect cannot be neglected. The total s-electron hyperfine anomaly may be written as (65)

$$^1\Delta_s^2 \cong {}^1\Delta_{BW}^2 + {}^1\Delta_{BR}^2 \tag{85}$$

where Δ_{BW} and Δ_{BR} are the isotopic differences of the Bohr-Weisskopf and the Breit-Rosenthal corrections, respectively. From the calculations of ROSENBERG and STROKE

Table 13. Hyperfine anomalies for some 4d and 5d elements

Isotopes	$^1\Delta^2_{s,exp}$ [%]	Reference	$^1\Delta^2_{s,calc}$ [%]
^{95,97}Mo	-0.0101 (2)	156	
^{107,109}Ag	-0.414 (1)	106	-0.369
^{185,187}Re	0.026 (5)	66	0.019
^{191,193}Ir	0.64 (7)	180	0.60
^{197,199}Au	3.66 (26)	106	3.50

/108/ who used a diffuse (Hofstaedter) nuclear charge distribution one finds $^{95}\Delta^{97}_{BR} \cong$ 0.007% and $^{185}\Delta^{187}_{BR} \cong$ 0.016%. Using (86) and $^1\Delta^2_{s,exp}$ from Table 13 leads to $^{95}\Delta^{97}_{BW} \cong$ -0.017% and $^{185}\Delta^{187}_{BW} \cong$ 0.010%. Unfortunately, there exists no estimate of the Bohr-Weisskopf correction for Mo using configuration mixing /112/ or core-polarization techniques /117/. For the deformed nuclei ^{185,187}Re the Bohr-Weisskopf correction has been evaluated /177/ using the procedure outlined by CLARK et al. /115/. Using Nilsson wavefunctions /218/ for a deformation parameter $\beta = 0.2$ and the quenched g factors g_s and the core g factors g_R from a compilation by BOHR and MOTTELSON /219/ one calculates $^{185}\Delta^{187}_{BW} \cong$ 0.003%. This value is too small, but of the right sign and the right order of magnitude.

8. Concluding Remarks

This study has dealt with atomic structure studies of the transition elements with
an unfilled 4d or 5d electron shell which have been performed mainly with the ABMR
technique. Although the agreement between theory and experiment is satisfactory in
principle, it has been shown how difficult it is to account in detail for the five
observables excitation energy, g_J value, magnetic dipole hfs constant A, electric
quadrupole hfs constant B, and level isotope shift for several states of an atom
simultaneously. Special emphasis has been laid on distinguishing between the effects of relativity and configuration interaction in hfs. For this purpose the experimental hfs data have been analyzed with respect to the effective operator formalism. This analysis has shown that the method of effective operators works quite
well, even in the region of heavy complex atoms. However, the hfs constants cannot
be reproduced within their experimental-error limits. In order to resolve these
discrepancies, it seems very desirable to improve the eigenvectors by taking into
account effective three-particle fine-structure interactions and to include SL-dependent configuration interaction effects in the hfs analysis which have been neglected so far.

The effective radial hyperfine integrals obtained from the analysis of the experimental hfs data have been compared with theoretical values calculated by relativistic self-consistent-field methods. This comparison clearly indicates the
strong influence of configuration interaction on the hfs of the 4d- and 5d-shell
atoms.

On the other hand, because the hyperfine interaction depends on properties both
of the electrons and of the nucleus of the atom under study, these extensive hfs
measurements have provided new information about the nuclear ground states of some
4d and 5d elements. Magnetic dipole moments, electric quadrupole moments, and hyperfine anomalies have been determined, constituting sensitive tests of nuclear models.

In the near future the use of tunable dye lasers in connection with ABMR experiments should lead to an extension of precise hyperfine and Zeeman studies to still
higher lying metastable states. If the hfs could be measured for enough states in
a single element, it might well be possible to achieve a much more comprehensive

understanding of atomic structure. In addition, using high-resolution Doppler-free laser spectroscopic methods extensive studies of isotope shifts and of the hfs of odd-parity atomic levels of the transition elements become possible.

The high efficiency and isotopic selectivity of laser-induced detection of metastable atoms might also be an interesting technique for ABMR experiments on radioactive isotopes.

Finally, the further development of multiconfiguration Dirac-Fock and many-body techniques for calculating configuration interaction effects can be expected to lead to improvements in the theoretical understanding of the atomic structure for complex atoms.

Acknowledgements. I greatly appreciate the cooperation of all members of the atomic beams group at Bonn.

I am grateful to Professor S. Penselin for his support and continuing interest in this work, and I wish to thank him and Dr. F. Träber for many helpful discussions during the preparation of this article and for a critical reading of the manuscript.

The financial support of the Deutsche Forschungsgemeinschaft is gratefully acknowledged. The numerical calculations were performed with the computer system IBM 370/168 of the Regionales Hochschulrechenzentrum der Universität Bonn.

References

1   C.M. Lederer, V.S. Shirley: *Table of Isotopes*, 7th ed. (Wiley, New York 1978) Appendix VII
2   F. Scheck: Phys. Rep. *44*, 189 (1978)
3   H. Engfer, H. Schneuwly, J.L. Vuilleumier, H.K. Walter, A. Zehnder: At. Data Nucl. Data Tables *14*, 509 (1974)
4   F. Boehm, P.L. Lee: At. Data Nucl. Data Tables *14*, 605 (1974)
5   C.W. de Jager, H. de Vries, C. de Vries: At. Data Nucl. Data Tables *14*, 479 (1974)
6   J.S.M. Harvey: Proc. R. Soc. London A *285*, 581 (1965)
7   P.G.H. Sandars, J. Beck: Proc. R. Soc. London A *289*, 97 (1965)
8   B.R. Judd: Proc. Phys. Soc. London *82*, 874 (1963)
9   B.R. Judd: In *La Structure Hyperfine Magnétique des Atomes et des Molécules*, ed. by R. Lefebvre, C. Moser (C.N.R.S., Paris 1967) p. 311
10 P.G.H. Sandars: Adv. Chem. Phys. *14*, 365 (1969)
11 I. Lindgren, A. Rosén: Case Stud. At. Phys. *4*, 93 (1974)
12 I. Lindgren, J. Morrison: *Atomic Many-Body Theory*, Springer Series in Chemical Physics, Vol. 13 (Springer, Berlin, Heidelberg, New York 1982)
13 J. Andriessen, M. Vajed-Samii, K. Raghunathan, S.N. Ray, T.P. Das: Hyperfine Interact. *4*, 91 (1978)
14 C. Froese-Fischer: *The Hartree-Fock Method for Atoms* (Wiley, New York 1977)
15 I.I. Rabi, J.R. Zacharias, S. Millman, P. Kusch: Phys. Rev. *53*, 318 (1938)
16 P. Kusch, S. Millman, I.I. Rabi: Phys. Rev. *57*, 765 (1940)
17 N.F. Ramsey: *Molecular Beams* (Oxford U.P., New York 1956)
18 H. Kopfermann: *Nuclear Moments* (Academic, New York 1958)
19 W.A. Nierenberg: Annu. Rev. Nucl. Sci. *7*, 349 (1957)
20 P. Kusch, V.W. Hughes: In "Atomic and Molecular Beam Spectroscopy", *Handbuch der Physik*, Vol. 37/1, ed. by S. Flügge (Springer, Berlin, Göttingen, Heidelberg 1959)
21 W.J. Childs: Case Stud. At. Phys. *3*, 215 (1973)
22 S. Penselin: In *Progress in Atomic Spectroscopy*, ed. by W. Hanle, H. Kleinpoppen (Plenum, New York 1978) Pt.A, p. 463
23 G.H. Fuller: J. Phys. Chem. Ref. Data *5*, 835 (1976)
24 The Göteborg-Uppsala Atomic-Beam Group: "Summary of Results from Measurements of Nuclear Spins and Moments" (1980) (unpublished)
25 C. Ekström, I. Lindgren: In *Atomic Physics 5*, ed. by R. Marrus, M. Prior, H. Shugart (Plenum, New York 1977) p. 201
26 G. zu Putlitz: "Determination of Nuclear Moments with Optical Double Resonance", in *Springer Tracts in Modern Physics*, Vol. 37 (Springer, Berlin, Heidelberg, New York 1965) p. 105
27 G.W. Series: Proc. 10th Scottish Universities Summer School in Physics (Academic, New York 1970) p. 395
28 W. Happer: Rev. Mod. Phys. *44*, 169 (1972)
29 W.J. Childs, B. Greenebaum: Phys. Rev. A *6*, 105 (1972)
30 U. Johann, J. Dembczyński, W. Ertmer: Z. Phys. A *303*, 7 (1981)
31 C. Bauche-Arnoult: Proc. R. Soc. London A *322*, 361 (1971)

32 C. Bauche-Arnoult: J. Phys. Paris *34*, 301 (1973)
33 S. Büttgenbach, G. Meisel, S. Penselin, K.H. Schneider: Z. Phys. *230*, 329 (1970)
34 C. Schwartz: Phys. Rev. *97*, 380 (1955)
35 A.R. Edmonds: *Drehimpulse in der Quantenmechanik* (Bibliographisches Institut, Mannheim 1964)
36 G. Breit: Phys. Rev. *34*, 553 (1929)
37 H.A. Bethe, E.E. Salpeter: *Quantum Mechanics of One- and Two-Electron Atoms* (Springer, Berlin, Göttingen, Heidelberg 1957)
38 A.I. Akhiezer, V.B. Berestetskii: *Quantum Electrodynamics* (Wiley-Interscience, New York 1965)
39 L. Armstrong, Jr.: *Theory of the Hyperfine Structure of Free Atoms* (Wiley-Interscience, New York 1971)
40 I.P. Grant: Adv. Phys. *19*, 747 (1970)
41 J.P. Desclaux: Phys. Scr. *21*, 436 (1980)
42 J.P. Desclaux: Comput. Phys. Commun. *9*, 31 (1975)
43 I.P. Grant, B.J. McKenzie, P.H. Norrington, D.F. Mayers, N.C. Pyper: Comput. Phys. Commun. *21*, 207 (1980)
44 J.P. Desclaux, C.M. Moser, G. Verhaegen: J. Phys. B *4*, 296 (1971)
45 H.J. Kluge, H. Sauter: Z. Phys. *270*, 295 (1974)
46 J. Bauche, G. Couarraze, J.J. Labarthe: Z. Phys. *270*, 311 (1974)
47 J.B. Mann, W.R. Johnson: Phys. Rev. A *4*, 41 (1971)
48 W. Buchmüller: Phys. Rev. A *18*, 1784 (1978)
49 J. Sucher: Phys. Rev. A *22*, 348 (1981)
50 K.A. Brueckner: Phys. Rev. *100*, 36 (1955)
51 J. Goldstone: Proc. R. Soc. London A *239*, 269 (1957)
52 H.P. Kelly: Phys. Rev. *131*, 684 (1963)
53 P.G.H. Sandars: In *La Structure Hyperfine Magnétique des Atomes et des Molécules*, ed. by R. Lefebvre, C. Moser (C.N.R.S., Paris 1967) p. 111
54 B.R. Judd: *Second Quantization and Atomic Spectroscopy* (The Johns Hopkins Press, Baltimore 1967)
55 B.R. Judd: Adv. Chem. Phys. *14*, 91 (1969)
56 I. Lindgren: J. Phys. B *7*, 2441 (1974)
57 J. Dembczyński: J. Phys. Paris *41*, 109 (1980)
58 U. Johann: Thesis, University of Bonn (1981)
59 R.M. Sternheimer: Phys. Rev. *80*, 102 (1950)
60 R.M. Sternheimer: Phys. Rev. *84*, 244 (1951)
61 R.M. Sternheimer, R.F. Peierls: Phys. Rev. A *3*, 837 (1971)
62 R.M. Sternheimer, R.F. Peierls: Phys. Rev. A *4*, 1722 (1971)
63 R.M. Sternheimer: Phys. Rev. A *6*, 1702 (1972)
64 I. Lindgren: In *Atomic Physics 4*, ed. by G. zu Putlitz, E.W. Weber, A. Winnacker (Plenum, New York 1975) p. 747
65 J. Bauche, C. Bauche-Arnoult: J. Phys. B *10*, L125 (1977)
66 K.H. Bürger, B. Burghardt, S. Büttgenbach, R. Harzer, H. Hoeffgen, G. Meisel, F. Träber: Z. Phys. A (1982) in press
67 I. Lindgren, S. Lundqvist (eds.): Proceedings of Nobel Symposium 46, Phys. Scr. *21*, No.3/4 (1980)
68 A. Hibbert: Rep. Prog. Phys. *38*, 1217 (1975)
69 S. Feneuille, L. Armstrong, Jr.: Phys. Rev. A *8*, 1173 (1973)
70 L. Armstrong, Jr., S. Feneuille: Adv. At. Molec. Phys. *10*, 1 (1974)
71 J. Andriessen, S.N. Ray, T. Lee, T.P. Das, D. Ikenberry: Phys. Rev. A *13*, 1669 (1976)
72 K. Raghunathan, J. Andriessen, S.N. Ray, T.P. Das: Hyperfine Interact. *4*, 96 (1978)
73 J. Andriessen: Thesis, University of Delft (1980)
74 R. Glass, A. Hibbert: Comput. Phys. Commun. *16*, 19 (1978)
75 C.G. Darwin: Proc. R. Soc. London A *118*, 654 (1928)
76 G. Racah: Phys. Rev. *63*, 367 (1943)

77 S. Feneuille: J. Phys. Paris *28*, 61 (1967)
78 E.U. Condon, G.H. Shortley: *The Theory of Atomic Spectra* (Cambridge U.P., Cambridge 1967)
79 G. Racah: Phys. Rev. *62*, 438 (1942)
80 G. Racah: Phys. Rev. *76*, 1352 (1949)
81 J.A. Barnes, B.L. Carroll, L.M. Flores: At. Data 2, 101 (1970)
82 C.W. Nielson, G.F. Koster: *Spectroscopic Coefficients for the p^n, d^n, and f^n Configurations* (MIT Press, Cambridge 1963)
83 B.G. Wybourne: *Spectroscopic Properties of Rare Earths* (Interscience, New York 1965)
84 S. Feneuille: J. Phys. Paris *28*, 497 (1967)
85 S. Feneuille: J. Phys. Paris *28*, 701 (1967)
86 J. Schrijver, P.E. Noorman: Physica *64*, 269 (1973)
87 J. Schrijver, P.E. Noorman: Physica *68*, 615 (1973)
88 K. Rajnak, B.G. Wybourne: Phys. Rev. *132*, 280 (1963)
89 S. Feneuille: J. Phys. Paris *28*, 315 (1967)
90 A. Pasternak, Z.B. Goldschmidt: Phys. Rev. A *6*, 55 (1972)
91 B.R. Judd: *Operator Techniques in Atomic Spectroscopy* (McGraw-Hill, New York 1963)
92 H.H. Marvin: Phys. Rev. *71*, 102 (1947)
93 L. Armstrong, Jr., S. Feneuille: Phys. Rev. *173*, 58 (1968)
94 W.J. Childs: Phys. Rev. A *2*, 316 (1970)
95 A. Abragam, J.H. van Vleck: Phys. Rev. *92*, 1448 (1953)
96 G. Breit: Nature *122*, 649 (1928)
97 H. Margenau: Phys. Rev. *57*, 383 (1940)
98 W. Perl: Phys. Rev. *91*, 852 (1953)
99 B.R. Judd, I. Lindgren: Phys. Rev. *122*, 1802 (1961)
100 M. Philips: Phys. Rev. *76*, 1803 (1949)
101 S. Büttgenbach, R. Dicke, H. Gebauer, M. Herschel: Internal Report, Institut für Angewandte Physik, Bonn (1976)
102 W.E. Lamb: Phys. Rev. *60*, 817 (1941)
103 F.D. Feiock, W.R. Johnson: Phys. Rev. *187*, 39 (1969)
104 G.K. Woodgate: Proc. R. Soc. London A *293*, 117 (1966)
105 H.H. Stroke: In *Atomic Physics*, ed. by B. Bederson, V.W. Cohen, F.M.J. Pichanick (Plenum, New York 1969) p. 523
106 G.H. Fuller, V.W. Cohen: Report ORNL-4591, Oak Ridge National Laboratory (1970)
107 E. Rosenthal, G. Breit: Phys. Rev. *41*, 459 (1932)
108 H.J. Rosenberg, H.H. Stroke: Phys. Rev. A *5*, 1992 (1972)
109 A. Bohr, V.F. Weisskopf: Phys. Rev. *77*, 94 (1950)
110 A. Bohr: Phys. Rev. *81*, 331 (1951)
111 A.S. Reiner: Nucl. Phys. *5*, 544 (1958)
112 H.H. Stroke, R.J. Blin-Stoyle, V. Jaccarino: Phys. Rev. *123*, 1326 (1961)
113 I. Unna: Phys. Lett. *24B*, 499 (1967)
114 G.J. Perlow: In *Hyperfine Interactions in Excited Nuclei*, ed. by G. Goldring, R. Kalish (Gordon and Breach, New York 1971) p. 651
115 D.L. Clark, M.E. Cage, D.A. Lewis, G.W. Greenlees: Phys. Rev. A *20*, 239 (1979)
116 P.A. Moskowitz, M. Lombardi: Phys. Lett. *46B*, 334 (1973)
117 T. Fujita, A. Arima: Nucl. Phys. A *254*, 513 (1975)
118 C. Ekström, L. Robertsson, S. Ingelman, G. Wannberg, I. Ragnarsson: Nucl. Phys. A *348*, 25 (1980)
119 S. Büttgenbach, R. Dicke, H. Gebauer, R. Kuhnen, F. Träber: Z. Phys. A *286*, 333 (1978)
120 P. A. Vanden Bout, V.J. Ehlers, W.A. Nierenberg, H.A. Shugart: Phys. Rev. *158*, 1078 (1967)
121 A.G. Blachman, D.A. Landman, A. Lurio: Phys. Rev. *150*, 59 (1966)
122 U. Brinkmann, A. Steudel, H. Walther: Z. Angew. Phys. *22*, 223 (1967)
123 W. Zeiske, G. Meisel, H. Gebauer, B. Hofer, W. Ertmer: Phys. Lett. *55A*, 405 (1976)

124 M. Gustavsson, I. Lindgren, G. Olsson, A. Rosén, S. Svanberg: Phys. Lett. *62A*, 250 (1977)
125 G. Gölz: Diplome Thesis, University of Bonn (1979)
126 A. Beckman, K.D. Böklen, D. Elke: Z. Phys. *270*, 173 (1974)
127 S. Büttgenbach: Diplome Thesis, University of Bonn (1970)
128 R. Schieder, Diplome Thesis, University of Bonn (1971)
129 S. Millman: Phys. Rev. *55*, 628 (1939)
130 K.D. Böklen: Z. Phys. *270*, 187 (1974)
131 D. Elke: Thesis, University of Bonn (1973)
132 H. Lew, E. Lipworth: In *Methods of Experimental Physics*, Vol. 4A, ed. by V.W. Hughes, H.L. Schultz (Academic, New York 1967) p. 155
133 G. Wolber, H. Figger, R.A. Haberstroh, S. Penselin: Phys. Lett. *29A*, 461 (1969)
134 G. Wolber, H. Figger, R.A. Haberstroh, S. Penselin: Z. Phys. *236*, 337 (1970)
135 R.E. Honig: RCA Rev. *23*, 567 (1962)
136 W.M. Doyle, R. Marrus: Nucl. Phys. *49*, 449 (1963)
137 R.G. Schlecht, M.B. White, D.W. McColm: Phys. Rev. *138*, B306 (1965)
138 L. Armstrong, Jr., R. Marrus: Phys. Rev. *138*, B310 (1965)
139 J.M. Pendlebury, D.B. Ring, K.F. Smith: In *La Structure Hyperfine Magnétique des Atomes et des Molécules*, ed. by R. Lefebvre, C. Moser (C.N.R.S., Paris 1967) p. 71
140 J.M. Pendlebury, D.B. Ring: J. Phys. B *5*, 368 (1972)
141 H. Rubinsztein, I. Lindgren, L. Lindström, H. Riedl, A. Rosén: Nucl. Instrum. Meth. *119*, 269 (1974)
142 H. Rubinsztein, M. Gustavsson: Phys. Lett. *58B*, 283 (1975)
143 H. Rubinsztein, M. Gustavsson: Phys. Scr. *28*, 209 (1978)
144 M. Dubke, W. Jitschin, G. Meisel, W.J. Childs: Phys. Lett. *65A*, 109 (1978)
145 W. Ertmer, B. Hofer: Z. Phys. A *276*, 9 (1976)
146 M. Gustavsson, G. Olsson, A. Rosén: Z. Phys. A *290*, 231 (1979)
147 P. Grundevik, M. Gustavsson, I. Lindgren, G. Olsson, L. Robertson, A. Rosén, S. Svanberg: Phys. Rev. Lett. *42*, 1528 (1979)
148 S.D. Rosner, R.A. Holt, T.D. Gaily: Phys. Rev. Lett. *35*, 785 (1975)
149 B. Burghardt, W. Jitschin, G. Meisel: Appl. Phys. *20*, 141 (1979)
150 W. Jitschin, G. Meisel: Appl. Phys. *19*, 181 (1979)
151 C.E. Moore: *Atomic Energy Levels* (U.S. Government Printing Office, Washington, D.C. 1971)
152 G. Fricke, H. Kopfermann, S. Penselin: Z. Phys. *154*, 218 (1959)
153 S. Büttgenbach, R. Dicke, H. Gebauer, R. Kuhnen, F. Träber: Z. Phys. A *286*, 125 (1978)
154 S. Büttgenbach, R. Dicke, H. Gebauer, M. Herschel, G. Meisel: Z. Phys. A *275*, 193 (1975)
155 M. Dubke: Diplome Thesis, University of Bonn (1977)
156 S. Büttgenbach, M. Herschel, G. Meisel, E. Schrödl, W. Witte, W.J. Childs: Z. Phys. *266*, 271 (1974)
157 G. Guthöhrlein: Private communication
158 D. Wendlandt, J. Bauche, P. Luc: J. Phys. B *10*, 1989 (1977)
159 S. Büttgenbach, M. Herschel, G. Meisel, E. Schrödl, W. Witte, W.J. Childs: Z. Phys. *269*, 189 (1974)
160 S. Büttgenbach, R. Dicke, H. Gebauer, M. Herschel: Z. Phys. A *280*, 217 (1977)
161 Y.W. Chan, W.J. Childs, L.S. Goodman: Phys. Rev. *173*, 107 (1968)
162 K.H. Chanappa, J.M. Pendlebury: Proc. Phys. Soc. London *86*, 1145 (1965)
163 H. Dahmen, S. Penselin: Z. Phys. *200*, 456 (1967)
164 W. Fischer, H. Hühnermann, E. Krüger: Z. Phys. *216*, 136 (1968)
165 H. Figger, G. Wolber: Z. Phys. *264*, 95 (1973)
166 J. Vergès, J.F. Wyart: Phys. Scr. *17*, 495 (1978)
167 I.J. Spalding, K.F. Smith: Proc. Phys. Soc. London *79*, 787 (1962)
168 A. Nunnemann, D. Zimmermann, P. Zimmermann: Z. Phys. A *290*, 123 (1979)
169 S. Büttgenbach, M. Herschel, G. Meisel, E. Schrödl, W. Witte: Z. Phys. *260*, 157 (1973)

170 S. Büttgenbach, R. Dicke, H. Gebauer: Phys. Lett. *62A*, 307 (1977)
171 S. Büttgenbach, G. Meisel: Z. Phys. *244*, 149 (1971)
172 K.H. Bürger, S. Büttgenbach, R. Dicke, H. Gebauer, R. Kuhnen, F. Träber:
     Z. Phys. A *298*, 159 (1980)
173 S. Büttgenbach, R. Dicke, H. Gebauer, R. Kuhnen, F. Träber: Z. Phys. A *283*,
     303 (1977)
174 R. Harzer: Thesis, University of Bonn (1981)
175 K. Murakawa: Phys. Rev. *110*, 393 (1958)
176 S. Büttgenbach, R. Dicke, F. Träber: Phys. Rev. A *19*, 1383 (1979)
177 S. Büttgenbach, R. Dicke, G. Gölz, F. Träber: Z. Phys. A *302*, 281 (1981)
178 B. Buchholz, H.D. Kronfeldt, R. Winkler: Physica *96C*, 297 (1979)
179 G. Himmel: Z. Phys. *211*, 68 (1968)
180 S. Büttgenbach, R. Dicke, H. Gebauer, R. Kuhnen, F. Träber: Z. Phys. A *286*,
     333 (1978)
181 F. Träber: Thesis, University of Bonn (1981)
182 G. Müller, R. Winkler: Z. Phys. A *273*, 313 (1975)
183 B. Greenebaum, W.J. Childs, L.S. Goodman: Bull. Am. Phys. Soc. *16*, 532 (1971)
184 S. Büttgenbach, N. Glaeser, B. Roski, F. Träber: To be published
185 A.G. Blachman, D.A. Landman, A. Lurio: Phys. Rev. *161*, 60 (1967)
186 W.J. Childs, L.S. Goodman: Phys. Rev. *141*, 176 (1966)
187 J.F. Wyart: Phys. Scr. *18*, 87 (1978)
188 S. Büttgenbach, R. Dicke: Z. Phys. A *275*, 197 (1975)
189 N. Glaeser: Diplome Thesis, University of Bonn (1982)
190 R.E. Trees: J. Opt. Soc. Am. *49*, 838 (1959)
191 S. Büttgenbach, F. Träber: To be published
192 S. Büttgenbach: Unpublished material
193 Y. Shadmi, E. Caspi: J. Res. Nat. Bur. Stand. *72A*, 757 (1968)
194 G.G. Gluck, Y. Bordarier, J. Bauche, T.A.M. van Kleef: Physica *30*, 2068 (1964)
195 J. Dembczyński: Private communication
196 S. Penselin: Z. Phys. *154*, 231 (1959)
197 S. Büttgenbach, R. Dicke, H. Gebauer: Phys. Lett. *58A*, 56 (1976)
198 G.J. Ritter: Phys. Rev. *126*, 240 (1962)
199 S. Büttgenbach, G. Meisel: Z. Phys. *250*, 57 (1972)
200 S. Büttgenbach, M. Herschel, G. Meisel, E. Schrödl, W. Witte, W.J. Childs:
     Z. Phys. *263*, 341 (1973)
201 W.J. Childs, L.S. Goodman: Bull. Am. Phys. Soc. *12*, 509 (1967)
202 J. Bauche: Physica *44*, 291 (1969)
203 J. Bauche, R.J. Champeau: Adv. At. Molec. Phys. *12*, 39 (1976)
204 H. Grethen, R. Winkler, J. Bauche: Physica *98C*, 222 (1980)
205 P.S. Bagus, B. Liu, H.F. Schäfer: Phys. Rev. A *2*, 555 (1970)
206 W.J. Childs: Phys. Rev. *160*, 9 (1967)
207 K.D. Böklen: Thesis, University of Bonn (1971)
208 S. Büttgenbach: Thesis, University of Bonn (1973)
209 W.A. Nierenberg, G.O. Brink: J. Phys. Radium *19*, 816 (1958)
210 P.G.H. Sandars, G.K. Woodgate: Nature *181*, 1395 (1958)
211 G.K. Woodgate: Proc. R. Soc. London A *293*, 117 (1966)
212 S. Büttgenbach, M. Herschel, G. Meisel, E. Schrödl, W. Witte: Phys. Lett.
     43B, 479 (1973)
213 W. Dey, P. Ebersold, H.J. Leisi, F. Scheck, H.K. Walter, A. Zehnder: Helv.
     Phys. Acta *47*, 93 (1974)
214 J. Konijn, W. van Doesburg, G.T. Ewan, T. Johansson, G. Tibell: Nucl. Phys. A
     360, 187 (1981)
215 L.A. Schaller, W. Dey: Helv. Phys. Acta *47*, 482 (1974)
216 R.J. Powers, P. Martin, G.H. Miller, R.E. Welsh, D.A. Jenkins: Nucl. Phys. A
     230, 413 (1974)
217 K.E.G. Löbner, M. Vetter, V. Hönig: Nucl. Data Tables A *7*, 495 (1970)
218 J.P. Davidson: *Collective Models of the Nucleus* (Academic, New York 1968)
219 A. Bohr, B.R. Mottelson: *Nuclear Structure*, Vol. 2 (Benjamin, Reading 1975) p. 303

Abbreviations in this Work

ABMR	atomic beam magnetic resonance
calc	calculated
CI	configuration interaction
cw	continuous wave
EAL	extended average level
eff	effective
exp	experimental
FWHM	full width half maximum
HF	Hartree-Fock
hfs	hyperfine structure
IS	isotope shift
LIF	laser-induced fluorescence
MBPT	many-body perturbation theory
MCDF	multiconfiguration Dirac-Fock
MCHF	multiconfiguration Hartree-Fock
mes	mesonic
NMR	nuclear magnetic resonance
RCF	relativistic correction factor
rel	relativistic
rf	radio-frequency
RHF	restricted Hartree-Fock
SCF	self-consistent field
SUHF	spin-unrestricted Hartree-Fock
UHF	unrestricted Hartree-Fock

Subject Index

I. Lindgren, J. Morrison

Atomic Many-Body Theory

1982. 96 figures. XIII, 469 pages. (Springer Series in Chemical Physics, Volume 13). ISBN 3-540-10504-2

The unified description of atomic theory provided in this book bridges the gap between elementary books on quantum mechanics and present-day research in the field. Angular-momentum theory and the Hartree-Fock model are developed systematically and then applied to a number of physical problems. The treatment of many-body theory which then follows is based on a general form of the Rayleigh-Schrödinger perturbation theory, applicable to open-shell as well as closed-shell systems. The presentation in the book is based largely on graphical methods. Angular momentum graphs are used to represent the coupling between the spin and orbital angular momenta of the electrons, and the different terms in the perturbation expansion are expressed by means of "Feynman-like" – or Goldstone – diagrams. These diagrams are evaluated using the angular-momentum graphs developed in the early part of the book. The formalism is applied to a number of problems in atomic physics, such as the electroncorrelation energy, the electrostatic term structure and the spinorbit and hyperfine interactions. The final chapter deals with the exp(S) or coupled-cluster formalism in the pair approximation, which appears to be the most promising approach for accurate calculations of the structure of real atomic and molecular systems.

I. I. Sobelman

Atomic Spectra and Radiative Transitions

1979. 21 figures, 46 tables. XII, 306 pages. (Springer Series in Chemical Physics, Volume 1). ISBN 3-540-09082-7

Contents: Elementary Information on Atomic Spectra: The Hydrogen Spectrum Systematics of the Spectra of Multielectron Atoms. Spectra of Multielectron Atoms. – Theory of Atomic Spectra: Angular Momenta. Systematics of the Levels of Multielectron Atoms. Hyperfine Structure of Spectral Lines. The Atom in an External Electric Field. The Atom in an External Magnetic Field. Radiative Transitions. – References. – List of Symbols. – Subject Index.

I. I. Sobelman, L. A. Vainshtein, E. A. Yukov

Excitation of Atoms and Broadening of Spectral Lines

1981. 34 figures, 40 tables. X, 315 pages. (Springer Series in Chemical Physics, Volume 7). ISBN 3-540-09890-9

Contents: Elementary Processes Giving Rise to Spectra. – Theory of Atomic Collisions. – Approximate Methods for Calculating Cross Sections. – Collisions Between Heavy Particles. – Some Problems of Excitation Kinetics. – Tables and Formulas for the Estimation of Effective Cross Sections. – Broadening of Spectral Lines. – References. – List of Symbols. – Subject Index. – Errata for volume 1 of this series.

Springer-Verlag
Berlin
Heidelberg
New York

H. Haken, H. C. Wolf

Atom- und Quantenphysik

Eine Einführung in die experimentellen und theoretischen Grundlagen
2., überarbeitete und erweiterte Auflage. 1983. 247 Abbildungen. Etwa 390 Seiten. ISBN 3-540-11897-7

Dynamics of Gas-Surface Interaction

Proceedings of the International School on Material Science and Technology Erice, Italy, July 1-15, 1981
Editors: **G. Benedek, U. Valbusa**
1982. 132 figures. XI, 282 pages
(Springer Series in Chemical Physics, Volume 21)
ISBN 3-540-11693-1

The recent great advances in molecular beam techniques, triggered off by aerospace research, has open new perspectives in surface physics. This book, originated by International School held at the Majorana Centre in Erice, offers an updated review of studies on the structure and dynamics of solid surfaces and on gas-surface interaction by means of molecular beams comparing them with other techniques such as LEED, neutron and optical spectroscopies.
Novel topics treated in the book are the spectroscopy of surface phonons by atom scattering, the investigation of inelastic bound-state resonances, the characterization of adsorbates by atom scattering and surface-enhanced Raman scattering, and the detection of charge-density waves at the surface of layered crystals. All physicists and chemists, graduate students included, involved in the study of gas-surface interactions will find this book useful.

C.P. Slichter

Principles of Magnetic Resonance

Corrected 2nd printing of the 2nd revised expanded edition.
1980. 115 figures, 9 tables. XII, 397 pages
(Springer Series in Solid-State Sciences, Volume 1)
ISBN 3-540-08476-2

"...an improved introductory book... Slichter, a professor of physics, and his graduate students... are responsible for numerous original and significant research contributions that appear in the book. The clarity and style in which the book is written reveals Slichter's research expertise and talent as an excellent teacher and expositor... The referencing is so good that certain new priorities in research contributions to nmr appear that were not previously obvious..."   *Physics Today*

Structure and Collisions of Ions and Atoms

Editor: **I. A. Sellin**
With contributions by numerous experts
1978. 157 figures, 17 tables. XI, 350 pages
(Topics in Current Physics, Volume 5)
ISBN 3-540-08576-9

Contents: Quantum Electrodynamics in Strong and Supercritical Fields. – Relativistic Effects in Highly Ionized Atoms. – Theory of Inelastic Atom-Atom Collisions. – Excitation in Energetic Ion-Atom Collisions Accompanied by Electron Emmision. – X-Ray Production in Heavy Ion-Atom-Collisions. – Extensions of Beam Foil Spectroscopy. – Atomic Collisions in Solids.

Springer-Verlag Berlin Heidelberg New York

W. Demtröder

Laser Spectroscopy

Basic Concepts and Instrumentation
2nd corrected printing. 1982. 431 figures.
Approx. 710 pages
(Springer Series in Chemical Physics, Volume 5)
ISBN 3-540-10343-0